# About Island Press

Since 1984, the nonprofit Island Press has been stimulating, shaping, and communicating the ideas that are essential for solving environmental problems worldwide. With more than 800 titles in print and some 40 new releases each year, we are the nation's leading publisher on environmental issues. We identify innovative thinkers and emerging trends in the environmental field. We work with world-renowned experts and authors to develop cross-disciplinary solutions to environmental challenges.

Island Press designs and implements coordinated book publication campaigns in order to communicate our critical messages in print, in person, and online using the latest technologies, programs, and the media. Our goal: to reach targeted audiences—scientists, policymakers, environmental advocates, the media, and concerned citizens—who can and will take action to protect the plants and animals that enrich our world, the ecosystems we need to survive, the water we drink, and the air we breathe.

Island Press gratefully acknowledges the support of its work by the Agua Fund, Inc., Annenberg Foundation, The Christensen Fund, The Nathan Cummings Foundation, The Geraldine R. Dodge Foundation, Doris Duke Charitable Foundation, The Educational Foundation of America, Betsy and Jesse Fink Foundation, The William and Flora Hewlett Foundation, The Kendeda Fund, The Forrest and Frances Lattner Foundation, The Andrew W. Mellon Foundation, The Curtis and Edith Munson Foundation, Oak Foundation, The Overbrook Foundation, the David and Lucile Packard Foundation, The Summit Fund of Washington, Trust for Architectural Easements, Wallace Global Fund, The Winslow Foundation, and other generous donors.

# CONSERVATION FOR A NEW GENERATION

# Conservation for a New Generation

## Redefining Natural Resources Management

EDITED BY
Richard L. Knight and Courtney White

Washington • Covelo • London

Library of Congress Cataloging-in-Publication Data
Conservation in transition : new approaches to natural resources management / edited by Richard L. Knight and Courtney White.
p. cm.
Includes bibliographical references and index.
ISBN-13: 978-1-59726-437-2 (cloth : alk. paper)
ISBN-10: 1-59726-437-7 (cloth : alk. paper)
ISBN-13: 978-1-59726-438-9 (pbk. : alk. paper)
ISBN-10: 1-59726-438-5 (pbk. : alk. paper) 1. Conservation of natuarl resources. I. Knight, Richard L. II. White, Joseph Courtney, 1960-
S936.C66 2008
333.72—dc22

2008008687

Printed on recycled, acid-free paper 

Manufactured in the United States of America

10 9 8 7 6 5 4 3 2 1

To Aldo Leopold, Wallace Stegner, and Wendell Berry,

and their belief that the answer is somewhere in the mix
of human and natural communities.

*Go and find it.*

# CONTENTS

# ACKNOWLEDGMENTS

The inspiration for this book comes from two sources. First, our love for the land and its life-sustaining virtues. When one realizes that our health, prosperity, and happiness come from the land, love and concern for its health inevitably follow. Second, our attempt to capture the new land-people relationships that define today's conservation derives from the cumulative debt we feel to the many students, conservation practitioners, farmers, ranchers, and foresters who have touched our lives, and from our gratitude for their devotion to land health. Among these people we certainly include our contributors. For no other reason than that they care about land and people did they add this extra burden to their already too busy lives. Thanks to these special people, and thanks to the land!

RLK wishes to acknowledge his obligations to those who have inspired him these many years. Though the list of his debts to others is long, he would be remiss not to mention his parents, Ken and Ruth Knight; Nina Leopold Bradley; Curt Meine, and his wife Heather. What defines them is what captures all that is good in humans.

CW would like to acknowledge everyone involved with the Quivira Coalition, whose dedication and good cheer have been the source of much of his inspiration. He acknowledges also his family—his other source of inspiration—his wife, Gen, and their children, Sterling and Olivia, whose future is a daily motivation.

Before our contributors submitted their essays we all assembled at Colorado State University to meet one another and listen to one another's comments. That meeting was essential to the book project, but it also allowed people from around the world to confirm their suspicions that conservation today is quite different from what it was only a decade ago. The meeting was made possible by the Warner College of Natural Resources at Colorado State University and by the Bradley Fund for the Environment of the Sand County Foundation. Kevin McAleese, representing the Sand County Foundation, was indispensable in planning the meeting. Mike Eckhoff ably assisted us in the conference organization.

Last, we take great pleasure in thanking our editors, Barbara Dean and Barbara Youngblood of Island Press. For readers contemplating their own book projects, here's an inside tip: Go with Island Press and just hope you get to work with Barbara and Barbara!

# INTRODUCTION

*Richard L. Knight*

I vividly recall a comment, nearly two decades ago, by a graduate student during a seminar at Colorado State University. Faculty and students were listening to a distinguished speaker explain the concepts behind the new discipline of conservation biology. Understandably, some faculty from my College of Natural Resources were muttering under their breath while the speaker extolled the virtues of this new entry to the field of conservation. In effect, my colleagues were saying, "Why do we need a new discipline when, after all, our own fields of wildlife, fishery, forestry, and range have been addressing similar issues for decades and are doing just fine, thank you." The student leaned over to me and whispered, "These disagreements are a problem of your generation, not mine."

What the student meant, of course, is that his generation, the next cohort of natural resource practitioners, was excited and invigorated by the flood of ideas coming from new disciplines entering the well-entrenched fields of natural resources management. If we, the faculty, wanted to argue the virtues and ills of those new disciplines, so be it—but his generation was moving ahead and welcomed the changes. I got it!

* * * * *

Two decades ago, the arena of natural resources management was confronted with a host of new disciplines (table I.1). Human dimensions, restoration ecology, landscape ecology, ecological economics, conservation biology, environmental ethics, geospatial sciences, and other new fields appeared, seemingly overnight, to expand the horizons of conservation practitioners and university programs.[1] What was different about those new disciplines was their metadisciplinary nature. They were more broadly based and more integrative than the traditional disciplines in natural resources. For example, wildlife biology had become focused on the population dynamics of single species—economically valuable species, endangered species, or species that threatened human economies. Conservation biology, on the other hand, took a more extensive view, from genes to organisms to biotic communities to ecosystems. Likewise, landscape ecology changed the way practitioners and academics viewed ecosystems, and conversations in natural resources management were now enriched by acknowledging the spatial and temporal dimensions of conservation work. Human dimensions balanced the one-sided emphasis on the "scientific management" of natural resources by admitting the importance of social capital and the multidimensional nature of our diverse publics. Ecological economics reminded us of the value of ecosystem services produced by healthy lands; by emphasizing the economic importance of restored landscapes, it provided conservationists with a powerful argument for conserving rather than developing land.

Natural resources management today is quite different from what it was decades earlier, and the changes have been remarkably well accepted. Students and faculty alike seem comfortable with the more integrative and crosscutting approaches. Practitioners also see the energy the new disciplines have brought to how we practice conservation. For example, fish and wildlife managers have benefited enormously from the ideas of conservation genetics, minimum viable populations, metapopulations, spatially explicit models, and new approaches in designing protected areas, all ideas that emanated from the field of conservation biology.

**Table I.1**
**Traditional and Emerging Disciplines in Conservation**

| Traditional Disciplines | Emerging Disciplines |
|---|---|
| Forestry science | Landscape ecology |
| Natural resource economics | Ecological economics |
| Fishery science | Conservation biology |
| Wildlife biology | Environmental ethics |
| Watershed science | Human dimensions in natural resources |
| Range science | Restoration ecology |
| Outdoor recreation | Geospatial sciences |

These new ideas have affected who is doing natural resources management, as well. Governmental organizations (GOs), from city open-space programs to state departments of natural resources to federal land-management services, now employ human dimension specialists and hire ecological economists. Nongovernmental organizations (NGOs) have taken the ideas of landscape ecology and conservation biology to new heights, designing and implementing conservation strategies at landscape and regional levels. It is safe to say that the number of job descriptions, as well as the number of actual jobs, in the broad field of conservation and natural resources management has increased severalfold.

In the dynamic landscape of contemporary conservation, all of those "emerging" disciplines are now well established; they have professional organizations, journals, annual meetings, and memberships that are outpacing those of the traditional disciplines. In Colorado State's Warner College of Natural Resources, the outdoor recreation department has changed its name to Human Dimensions of Natural Resources, while the fish and wildlife biology department is now known as the Department of Fish, Wildlife, and Conservation Biology. Likewise, the forest, range, and watershed programs have merged into a Department of Forest, Rangeland,

and Watershed Stewardship with a robust emphasis on restoring degraded watersheds and promoting healthy natural and human communities. That student who spoke to me two decades ago was right: Change empowered by the entry of new ideas *was* an issue of my generation, for what seemed like radical ideas then are today universally embraced.

As natural resources management was experiencing profound changes in its intellectual repertoire, our home planet and its inhabitants were likewise showing significant alterations, ranging from destabilization of the atmosphere to a global extinction spasm. And, because of our burgeoning human population, inappropriate technologies, and rising materialism, the earth is increasingly in a degraded state.[2] The changes in the health of the only planet we will ever know have hastened the acknowledgment that land uses that decrease land health are nonsustainable, as are human communities dependent on those uses. Collectively, those changes reveal the obvious: that human communities cannot survive and prosper on degraded lands, that land health and human health are entwined, indivisible. Wendell Berry sank the stake when he wrote, "You cannot save the land apart from the people or the people apart from the land. To save either, you must save both."[3]

What is needed now is not to revisit those "new" disciplines, for they have become fully integrated into the traditional fields of natural resources management. What is different from a decade or so ago is how the conservation of natural resources is being *practiced*.

Inspired by Aldo Leopold's writings,[4] this new way of managing natural resources is captured by the following comment: "Conservation occurs at the nexus of land use and land health. Therefore, conservation that works is conservation that works well for both people and land. Actions that benefit one at the expense of the other are not conservation, they are something else. Healthy natural communities coincide with healthy human communities; the reverse seldom occurs."[5] This new way of doing conservation is not opposed to human land uses, whether they be logging or outdoor recreation. It is opposed to land uses that degrade the health of the land (in the Leopoldian sense of "land health"[6]). In other words, natural resources management in the twenty-first century is all about land health and prosperous human communities.

Historically, natural resources conservation was the result of federal initiatives. Legislation was passed, appropriations provided, and new programs implemented with a top-down approach.[7] Today conservation is often the result of bottom-up initiatives emanating from watersheds and supported by local governments, tax incentives, and financial support from private entities and NGOs (table I.2). In the face of rapid change, institutions, agencies, universities, and practitioners are experimenting with new approaches to natural resources management. On the ground, practitioners are reexamining the role of institutions and agencies and trying new forms of governance. In universities, teachers and students are placing emphasis on topics such as human dimensions and valuing ecosystem services. The private sector, whether NGOs or foundations with money to support conservation, is becoming increasingly important in shaping the landscape of conservation. The traditional focus on wilderness protection is being augmented with the emerging idea of working wildlands—that is, keeping lands productive while keeping human densities low, thereby maintaining the capacity of the land to support biodiversity.

Although there are many defining elements of this new strain of natural resources management, they all have in common the following elements that revolve around land, people, institutions and organizations, ecology, and economics:

- They work across administrative boundaries rather than staying within them.
- They integrate social capital with ecological and economic dimensions.

**Table I.2**
**Contrasting Approaches to Conservation**

| Top-Down | Bottom-Up |
|---|---|
| Land use | Land health |
| Litigation | Cooperation |
| Federal monies and unfunded mandates | Economic incentives and private support |
| Public land | Private-public land |
| Wilderness | Working wildlands and landscape linkages |
| Working within administrative boundaries | Working across administrative boundaries |
| Command and control | Adaptive management |
| Single species | Ecological processes and keystone species |
| Disciplinary focus | Metadisciplinary approach |
| Technical expertise | Social capital |

*Note:* These approaches are best viewed as a continuum rather than hard and fast dichotomous categories.

- They encourage bottom-up participation rather than strictly top-down initiatives.
- They acknowledge all biodiversity rather than only economically valuable species.
- They emphasize economic incentives rather than federal appropriations and unfunded mandates.
- Finally, they are not exclusionary; rather, they are inclusionary. All interests are welcome at the table, not just favorites at any particular time. Translated, this means that all appropriate land uses are encouraged but, importantly, only to the degree that the land can sustainably accommodate those uses.

Empowered by these concepts, hundreds of attempts to reinvent natural resources management have been occurring at the watershed level nationwide.[8] A defining component of most of those efforts is that they have been locally initiated and are more bottom up than top down. Another component they all share is that few of them were initiated by federal agencies. Indeed, most federal, top-down efforts are increasingly sharing their "authority of the resource" with more local agencies (state, county, city) and NGOs (and, interestingly, rediscovering the ideas of early progressive thinkers such as Lewis Mumford).[9]

Just as ecosystem management has never been codified by Congress, the attempts taking place across the United States to build healthy human and natural communities are, at best, only loosely structured by mandates from the Beltway. It is time to capture the entrepreneurial strain that is providing the momentum around natural resources management in today's changing world. It is time to coalesce this vast amount of tinkering and experimentation into a coherent blueprint for conservation that works. In other words, now that conservation is being increasingly shaped by local initiatives, it is time to examine whether this new model of conservation, which is more regional and bottom up, is compatible with our equally fast-shifting society and changing environment.

Our goal in this multiauthored book is to capture the dynamic state of natural resources management in the years following the rise of ecosystem management in the early 1990s. Almost every component of land–land use–people relationships is in flux, and no one has yet attempted to capture the change. The contributors to this book are actively engaged in these new approaches. For every new approach to conservation described in the chapters of this book, there is a case study by a practitioner highlighting how the approach is working on the ground.

Our book consists of three parts bracketed by opening and closing chapters that explore the fundamental truth that unites human and land communities: As one prospers, so does the other; as one declines, so, too, will the other. In the opening chapter Curt Meine provides a narrative of human health and land health through the eyes of a Cree Indian tribe north of the Arctic Circle. Woven into his story is a capsule of conservation and the stages it encompasses, beginning with unregulated exploitation of natural resources and moving to the development of conservation and preservation, followed by environmentalism, and ending with the current paradigm of ecosystem management.

Part I of our book, "Agencies and Institutions: The Need for Innovation," has three chapters and four case studies. The chapter authors critically examine the need for the government agencies that manage our public lands to partner with diverse stakeholders, NGOs, and other components of the private sector in finding ways to promote healthy landscapes along with prosperous human communities. Building on the need for agencies to take a fresh approach to natural resources management, Part II explores emerging tools in the new toolbox of conservation. Titled "A Changing Toolbox for Conservation," it has four chapters and three case studies. Ranging from community planning to conservation planning, valuing ecosystem services, and the need to use economic incentives that promote conservation, the topics of these chapters, along with the case studies that illustrate the effectiveness of these new tools, demonstrate how natural resources management is changing. Part III, "The Radical Center: Finding Common Ground," consists of three chapters and five case studies. The chapters demonstrate the need for cooperation on natural resources issues that have historically divided us. The case studies illustrate how the emphasis on the importance of building strong relationships in natural resources management usually results in desired outcomes for both people and the land. Following Part III, the book ends with a chapter by Bob Budd and a conclusion by Courtney White. Budd tells a story of people and place, the North Piney River in Wyoming, to illustrate the connections between people who come to know and understand a watershed and the resulting solicitude they show for the land's health. Courtney White's conclusion provides a synopsis of the book, as well as a peek into the future, acknowledging the uncertainty that awaits us all.

* * * * *

What conservation is today is highly debatable. What is not debatable is that effective conservation—today and in the future—will be different from what it has been. To move successfully into our future, we need a vision that configures the changes into a comprehensible pattern. Such a blueprint not only will provide guidance to agencies and organizations, but also, and even more critically, will offer a conceptual and pragmatic road map for students, who are the future practitioners. Without such direction, it is certain that their education will prepare them for a time that has passed. The hope of this book is to provide that conceptual and pragmatic vision.

**Notes**

1. R. Knight and S. Bates, eds., *A New Century for Natural Resources Management* (Washington, DC: Island Press, 1995).
2. United Nations, *Millennium Ecosystem Assessment* (New York: United Nations, 2005).
3. W. Berry, "Private Property and the Common Wealth," in *Another Turn of the Crank* (Washington, DC: Counterpoint, 1995), p. 56.
4. C. Meine and R. Knight, eds., *The Essential Aldo Leopold: Quotations and Commentaries* (Madison: University of Wisconsin Press, 1999).
5. R. Knight, "Bridging the Great Divide: Reconnecting Rural and Urban Communities in the New West," in *Home Land: Ranching and a West that Works*, ed. L. Pritchett et al. (Boulder, CO: Johnson Books, 2007), pp. 13–25.
6. B. Callicott and E. Freyfogle, eds., *Aldo Leopold: For the Health of the Land* (Washington, DC: Island Press, 1999).
7. R. Nelson, "The Federal Land Management Agencies," in *A New Century for Natural Resources Management*, ed. R. Knight and S. Bates (Washington, DC: Island Press, 1995), pp. 37–60.

8. S. Yaffee et al., *Ecosystem Management in the United States: An Assessment of Current Experience* (Washington, DC: Island Press, 1996); G. Meffe et al., *Ecosystem Management: Adaptive, Community-Based Conservation* (Washington, DC: Island Press, 2002).
9. B. Minteer, "Regional Planning as Pragmatic Conservation," in *Reconstructing Conservation: Finding Common Ground*, ed. B. Minteer and R. Manning (Washington, DC: Island Press, 2003), pp. 93–113.

CHAPTER ONE

# THIS PLACE IN TIME

*Curt Meine*

We head north at sunset through choppy waters along the east shore of James Bay. Fred guides our fleet of three fully loaded, twenty-foot freighter canoes though a labyrinth of islands, mainland points, and submerged granite ledges. Fred is the *ouchimaw* in this part of the Cree nation of Wemindji. Among the James Bay Cree, the *ouchimawch* serve (in the words of one student of their vital role) as "senior grassroots managers of this vulnerable ecosystem."[1]

We have spent several days at the community's annual gathering on Old Factory Island, forty miles downshore as the canoe glides (figure 1.1). Now we are heading back to the village, where the Cree relocated two generations ago, in 1959. Bouncing over the waves in the pink subarctic twilight, we pass islands crowned in dark spires of white spruce and balsam fir. One small island catches my eye. Beneath a rise of barren granite, a series of terraces steps down toward the chilly waters of James Bay. The land around the bay in Quebec and Ontario is rebounding. At the time of the last glacial maximum, twenty thousand years ago, this place lay buried under five thousand meters of glacial ice at the heart of the Laurentide Ice Sheet. The burden was so immense that it compressed the earth's crust. Over the millennia, as the great ice sheet melted back, the depressed land

Figure 1.1 Old Factory Island, James Bay, Canada.

has sprung back. It is still rising. The geologist's term for the phenomenon is *isostatic rebound*.[2] The Cree speak of "the growing land."

The terraces on the small island are ancient beach ridges, each one marking a pause in time as the land has grown. The terrain at Wemindji has risen about seventy meters over the last six thousand years. It continues to rebound at the rate of about a meter per century—fast enough to outpace the rate of sea level rise that also came with the melting, fast enough even to be noted across a human lifetime. Wemindji's elders can tell you of places that have emerged from the waters, of plants and animals living differently here than they once did, of the Cree responding and adjusting.

My opportunity to be here has come through colleagues from McGill University in Montreal who have joined the Cree in an innovative partnership.[3] The academics and the Wemindji Cree are collaborating on a proposal to establish a protected area that would embrace two entire watersheds feeding into James Bay. The proposed protected area would coincide closely with the hunting territory that Fred oversees in his capacity as *ouchimaw*. It is a creative proposal that defies traditional expectations—as well as recent criticisms—of protected areas as a conservation strategy.[4]

The twelve hundred Cree of Wemindji represent the latest generation to live upon, and with, the growing land. By almost any conservation standard they have lived well here and have done so for some five thousand years. Cree traditions and practices have served to reinforce a tight network of reciprocal relationships connecting the land, the water, the plants and animals, the people, and the spirit.

In the four centuries since the arrival of the Shaped-Wood People from Europe, the resilience of those relationships has been constantly tested.[5] Yet, even against the backdrop of those last four centuries, the rate of change in the last two generations stands out as remarkable. Transformation has come to the culture, economy, and landscape of the James Bay Cree in a series of cascades, one consequence after another: the forced relocation of Cree children to government-supported residential schools;[6] the movement toward permanent settlement in Wemindji; the loss of the age-old pattern of families living a subsistence life in "the bush" for half the year; the announcement in 1971 of the Quebec provincial government's vast plan for hydropower development in the Cree lands east of James Bay; construction of the paved Route de la Baie James to facilitate the hydropower plan. Now the pressure to open gold and diamond mines in Wemindji country is growing. And in this subarctic land, the impacts of global warming on the ice and wind, the plants and animals are noticed even by the younger Cree. The Wemindji Cree wonder, along with communities around the world, how changes in their land will result in changes in their identity—and vice versa—and how they ought to respond.

My academic colleagues and my new friends from Wemindji are gathered to review the progress of their partnership. My appointed task, the reason for my even being here, is to offer a few relevant—I hope—words about Aldo Leopold and the land ethic in the land of the Cree. I am not at all convinced that this is possible.

The sun sets by the time Fred maneuvers our big canoe around the last spruce-studded point, into calmer waters, toward the lights of Wemindji.

* * * * *

As we may learn from the growing land, the *terra* is only relatively *firma*. Our science and our stories tell us that land changes and that human communities change. They change in different ways, at different rates. They change in response to each other. They change due to forces large and small, long-term and immediate, far away and close at hand. Amid such change, conservation aims to encourage ways of living by which we can meet our material needs, allow ourselves and our communities to flourish, express our human hopes, honor the beauty and mystery of the world, sustain its biological diversity, and promote its ecological health.

These are complex and interrelated aspirations. In pursuing them, conservationists have had to change, as the movement that first fully emerged a century ago has itself continually evolved. The story of conservation is one of shifting philosophical foundations, increasing scientific and historical knowledge, evolving public policy, and novel tools and techniques—all in dynamic interplay, occurring within a larger world of relentless cultural, economic, and environmental change.[7] To gain perspective as conservationists on our own place in time is no simple matter.

Conservation in its modern sense gained legitimacy and definition in the early 1900s, in the wake of an unprecedented, three-decade wave of private exploitation of North America's forests, prairies, rangelands, fisheries, and game populations. As conservation became official government policy under the leadership of President Theodore Roosevelt and his "chief forester" Gifford Pinchot, the utilitarian definition of conservation as the "wise use" of natural resources held sway. That definition carried corollaries: Conservation ought to serve "the greatest good for the greatest number over the long run"; it aimed to produce sustained yields of particular commodities (timber, water, fish, forage, game); it would achieve those sustained yields through efficient, professional, scientific management; it would strive to ensure fair distribution of the wealth that flowed from resource development. Conservation was conceived with the Progressive Era's faith in the capacity of science, technology, economics, and government to correct the ills wrought by the unbridled abuse of natural resources.

Meanwhile, walking with Pinchot but whispering into Roosevelt's other ear was their contemporary John Muir, the voice of a wilder America, of the big trees and monumental landscapes, of that strain of conservationist that sought to protect the beautiful, the unique, and the sublime. The preservationists could make common cause with the utilitarians, sharing as they did an appreciation of science and a faith in government's potential for effective administration. But they parted ways when "wise use" undermined the aesthetic, restorative, and spiritual values of wild things and wild places. The friction between utilitarians and preservationists would provide the dramatic storyline for much of conservation's long political drama across the twentieth century.

Aldo Leopold and his generation of conservationists inherited this philosophical tension in the 1930s and 1940s.[8] It drove Leopold's own conceptual innovations and evoked his plea for a unifying land ethic.[9] He could not abide merely material definitions of progress or the economic determinism and "ruthless utilitarianism" that in his view had disfigured the American landscape and revealed flaws in the character of American culture. Neither could he abide that approach to conservation that segregated aesthetics, averted its eyes from unpleasant economic reality, and sought refuge in the "parlor of scenic beauty."[10] For all of its success in establishing itself in the public mind and in government agencies, conservation had made scant progress toward reconciling its own multiple aims and achieving a more "harmonious balanced system of land use."[11]

Leopold once noted that our advanced technologies served to "crack the atom, to command the tides." "But," he continued, "they do not suffice for the oldest task in human history: to live on a piece of land without spoiling it."[12] That ultimate task could not be achieved simply by gaining new scientific knowledge or developing new tools; it required an *ethic* to better guide application of that knowledge and use of those tools. That ethic would avail itself of new ecological understanding and encourage new ways of valuing the nonhuman world. It would see land not merely as a commodity belonging to us, but "as a community to which we belong."[13] It recognized the interwoven history and destiny of people and land while demonstrating broad commitment to the common good. It called upon us all—as individuals and communities, producers and consumers, business owners and land managers, citizens and elected officials—to assume responsibility for the overall health of the land.[14]

Leopold gathered these ideas in "The Land Ethic," the capstone essay from his classic 1949 book *A Sand County Almanac*. The essay represented a bold advance in conservation, a leap beyond both the rationale and the tensions that had marked the young movement. In it, Leopold distilled the lessons acquired by an entire generation of conservationists. They had witnessed (and inevitably participated in) the mechanization and industrialization of the landscape; weathered the Dust Bowl, the Great Depression, and World War II; developed whole new fields and disciplines (including soil conservation, range management, and wildlife management); and, for the first time, brought findings from the emerging science of ecology into conservation practice.

"The Land Ethic" also anticipated the changes that would come with the rise of the environmental movement in the 1960s and 1970s. With its view of land not merely as a commodity but as "a community to which we belong" and its call to sustain "the integrity, stability, and beauty" of that community, "The Land Ethic" became a touchstone for subsequent generations of professional resource managers, landowners, and citizens alike as they confronted profound changes in social and environmental conditions.

Conservation evolved in divergent ways after World War II. On the one hand, it became a worldwide concern, and its professional ranks swelled. It absorbed revolutionary scientific findings in fields ranging from paleontology and geology to ecology and genetics. Its diagnostic and information technologies grew vastly in sophistication. It began to address a suite of concerns that accompanied the prosperity of the postwar years: the accelerated loss of wildlands; a growing list of threatened and endangered species; nuclear proliferation and the threat of atomic warfare; air and water pollution and other forms of environmental contamination; the development and indiscriminate use of new artificial chemical compounds; the triumph of the automobile culture; and the spread of suburbia.

On the other hand, conservation's ability to integrate new information and ideas, and to respond effectively to new threats, suffered in the postwar years. The conservation movement, as such, was fragmented among publics interested in various parts of the land (fish, game, soils, scenery, forests, rangelands, parks, rivers, wilderness, trails, etc.). Professional resource managers became increasingly specialized and focused on the output of their particular commodities (sport fish, game animals, commodity crops, visitor days, board feet, livestock forage, acre feet, kilowatts, etc.). The distinctions of *place* were overwhelmed by the need to get out the cut, meet the demand, enhance the visitor experience, maximize the output. Progress was measured according to raw economic and political benchmarks rather than any ethical one.

Along the way, conservation gave way to environmentalism. Environmentalism in the United States came of age along with the postwar, largely urban and suburban, baby-boom generation. Some older conservationists became environmentalists; others did not. Some younger environmentalists appreciated the strengths (as well as the shortcomings) of the older conservation tradition; others did not. In any case, the transition from the older conservation movement to modern environmentalism left behind it a tumultuous wake washing up against a complex and rocky shoreline. We are still riding over the roiling waves, in heavily laden canoes.

Yet, in the aftermath of the environmental awakening of the 1970s, creative conservationists began to challenge the fragmented, overspecialized, output-driven approaches that dominated professional natural resources management through much of the twentieth century. Over the last three decades, conservationists have gone about exploring, and inventing, a new approach, one that seeks safe passage through the shoals of harsh political ideology, toward common ground. And in seeking to restore and sustain healthier connections between people and land, conservationists have contributed importantly to a still broader societal need: reclaiming the vitality of community life.

The experiment in partnership growing at Wemindji is but one example of this flowering. We can also find such innovation in the more than two hundred institutions and organizations that belong to Chicago Wilderness, an extraordinary regional consortium that is redefining the notion of conservation in urban settings.[15] We find it in the Quivira Coalition's conventional wisdom-defying efforts to bring western ranchers and diverse environmental interests together in the shared pursuit of land stewardship and sustainable communities.[16] We find it in charismatic mega-landscapes, through the work of groups like the Greater Yellowstone Coalition and the Blackfoot Challenge.[17] We find it as well in less celebrated places, through a vast, proliferating array of conservancies, alliances, land trusts, initiatives, coalitions, networks, projects, partnerships, councils, and collaboratives.

This still emerging approach seeks to work fluently across multiple spatial scales, from the local to the regional to the global. It seeks also to be aware of multiple time scales, seeing its goals through varied layers of time and aiming to harmonize present and long-term needs. It sees land not as a collection of discrete parts, but as a complex, changing whole. It recognizes the need to work across disciplinary and jurisdictional boundaries and across entire landscapes. It appreciates degrees of human impact on the land and the intricate interrelationships between the natural and the cultural in any landscape. It aspires to sustain not merely *yields* but also the ecological functioning that underlies and defines land health. It recognizes the need to extend Leopold's notion of the land ethic to embrace aquatic and marine environments. It honors the bonds that link land to community life, economy, and identity. It acknowledges a hard reality: that all our places are weighed down by the overburden of past injustice and injury. It holds that the sustainability of human communities and economies cannot be defined apart from the land. It understands the necessity of community-based and participatory approaches in building the social foundations for effective conservation.[18]

These are hopeful signs, indications that, a century into the conservation-environmental movement, we might finally be getting the *scale* part, the *relationship* part, and the *integration* part of conservation right.

Although this movement-with-a-movement has been growing rapidly, its roots in conservation history reach deep. The standard narrative of environmental history has emphasized the expanding role of governmental agencies, wielding the stick of regulation and protecting public lands and other common resources through legislation and top-down management strategies. That narrative, however, misses the quieter, parallel development over the last century of progressive, participatory approaches to conservation based in local places and dedicated to what we would now call the sustainability of ecosystems, landscapes, and human communities.[19] It is a rich history that begs to be reclaimed. Examples of such approaches can be found in the following categories:

- *Watersheds.* Awareness of the fundamental significance of watersheds—as a geographical reality and as an organizing concept—dates to the very origins of the U.S. conservation movement. In his 1864 classic *Man and Nature*, George Perkins Marsh warned of "the great, the irreparable, the appalling mischiefs" that occur at the watershed scale due to destructive land clearing and ignorance of basic hydrological processes.[20] Marsh's work provided the rationale for whole fields of later conservation innovation: the landmark designation of the water-conserving forestland of New York's Adirondack Mountains in the 1870s and 1880s; John Wesley Powell's radical (and ill-fated) proposals in the 1890s to organize development of the American West along watershed lines; the designation of the early national forest reserves in the 1890s and 1900s to protect the headwaters of western streams. We can look back now upon the cooperative watershed restoration projects that began in the early 1930s as important antecedents of contemporary watershed- and community-based conservation projects.[21]
- *Land trusts.* As a force for private land conservation, the land trust movement in the United States has expanded phenomenally in the last several decades, with more than 1,700 land trusts now established across the nation. As a tool for land conservation, the land trust has been around for more than a century (with even older precedents in England and elsewhere). In 1891 civic leaders in Massachusetts advocated establishment of, and the state legislature chartered, the Trustees of Reservations. It was "the first private organization that included the essential features of a land trust: a mission dedicated to acquiring, holding, and maintaining natural, scenic, and historic sites."[22] The idea spread slowly in the decades that followed, with the incorporation of such small local organizations as the Connecticut Forest and Park Association (1895) and the Society for the Protection of New

Hampshire Forests (1901). The full potential of the land trust concept would be realized only after World War II, with the creation in 1946 of The Nature Conservancy as the first national land trust.[23]

- *Cooperative resource management.* When Garrett Hardin published his classic essay "The Tragedy of the Commons" in 1968, he focused attention on the inherent challenge of managing common resources.[24] Often overlooked in the discussion that has ensued ever since are the precedents for cooperative management of natural resources that lie deep in the history of conservation. Indeed, they predate the conservation movement per se and include indigenous stewardship practices from cultures around the globe. In Europe, the establishment of town forests dates to the Middle Ages. In New England, town forests were created as early as 1711 (in New Hampshire) and were established in state law throughout the region by the early 1900s. In the early decades of the twentieth century, cooperative game protection associations and experimental wildlife management partnerships emerged as landowners and sportsmen-conservationists worked together to restore wildlife populations and habitats.[25] At the other end of the spatial scale, we can find such examples as the complex efforts of state, provincial, tribal, and national governments to comanage the Great Lakes, which date to the early 1900s.[26]
- *Ecosystem management.* The emergence of ecosystem management in the late 1980s and early 1990s was taken by some to be a radical and unnecessary departure from traditional, practical approaches to the management of natural resources. Others saw it as a false front for even more egregious expression of those hubristic, heavy-handed management approaches. An alternative view, however, recognized in ecosystem management at least the potential to reclaim the legacy of a better integrated, ecologically informed, and historically grounded stewardship of the *land* (in Aldo Leopold's sense of the term—the

> entire community of "soils, waters, fauna, and flora, as well as people").[27] Lost in the post–World War II trend toward intensified, specialized resource development, that legacy from the 1930s and 1940s began to reemerge in the 1980s under the more scientifically sophisticated, globally relevant rubric of "sustainability." That dovetailed with the growing appreciation of conservation's intrinsic cultural dimension and the vital role of people and human communities in attaining—or undermining—conservation goals.

The evolution of conservation approaches, then, is a much more complicated story than we have generally accounted for. It is not a simple story of heavy-handed, top-down, command-and-control, governmental coercion; nor is it a simple story of progressive action realized solely through legislative initiative and legal decision. It is a more interesting story of constant change, occurring in different ways to meet different ends, involving a rich intermingling of social, economic, cultural, and environmental needs and hopes.

The evolution of conservation, of course, continues. It does so now amid sobering realities—realities that are converging in unprecedented ways. The greater context in which conservation exists is changing. In hindsight, we can view the conservation movement as the twin companion of the industrial age: It has grown as an idea, and a force for change and adjustment, within a time of cheap and abundant fossil fuels. We know that those energy sources are finite and that we may have already passed the point of peak oil. We know, too, that the exploitation of fossil fuels has had unintended consequences. Global warming is being experienced most directly by the land and creatures and people of the high latitudes (the Wemindji Cree among them), but no community on the planet will be immune to its impacts. To these forces of change we can add other global mega-trends: human population growth; the ever-increasing mobility and consumption of that human population; the spread of invasive species; the erosion of biological diversity; the increased incidence of emerging diseases; the rising demand for and consumption of freshwater; the degradation of the world's coastal and marine ecosystems. There was never a time like the one we have entered. If conservation is to answer effectively the call of these times, it will require unprecedented leadership, communication, and commitment.

It begins with us making connections; finding common ground; freeing our imaginations; and inventing new ways of thinking, being, and caring in the world.

It begins with us, here, now.

* * * * *

When I was first invited to Wemindji, I learned a Cree word, *Iyiyuuschii*, whose meaning I could not grasp. As I listened to those at Wemindji, it gained definition. I began to translate it in my own mind as "the land of the people." But such a word cannot be simply construed. As explained by the students working on the project, "Depending on the context, *iyiyuu* can represent the Cree people, all humans, or all life. *Ischii* can represent a hunting territory, Cree territory, the entire earth, or the soil. Together, *iyiyuu ischii* signifies interacting life on the land, with the perspective that the Cree people are not separate from the living land and their territory is not separate from the earth as a whole."[28]

What does this mean for my task at hand? What words can I possibly offer my Cree hosts? Leopold's land ethic is a complicated notion, representing a rich melding of western scientific knowledge, evolving ethical standards, and shifting definitions of community. Leopold offered his land ethic a little more than half a century ago. The James Bay Cree have had five millennia to distill their land wisdom. It is embedded in their *aatiyuuhkhn*, their sacred stories:

> *It is the responsibility of all peoples to protect and preserve the land.*
>
> *Water is sacred and is life-giving.*

*All peoples must live in harmony with the natural order.*

*There is medicine on the land, where beauty and strength can be found.*

*Knowledge comes from the Creator through the land.*

*Life is tied to and connected to the land.*

*Peoples must acknowledge, give thanks for, and honor what is received and taken from the land.*

*Everything has a spirit.*

*Eeyouch and all the peoples of the world are connected to the spirit of the land.*[29]

The Cree have endured upon this rich substrate of understanding. What can I possibly say, here, now, in *Iyiyuuschii*?

I begin by expressing my gratitude for the opportunity to be here and by invoking the great glaciers that once connected Wisconsin and Wemindji. Dorothy Stewart, who serves as the team's community liaison, steps up to translate my remarks for the benefit of the Cree elders and other community members. I hadn't thought about that! Immediately, I make adjustments. To allow time for translation, I cut my planned remarks in half. In fact, I pretty much dispense with my plan. This is not about explaining. This is about connecting.

I speak of geese. I was by now familiar with the vital role of the Canada goose in Cree culture. The goose has been described as a "cultural keystone species" for the James Bay Cree. The tails of the Air Creebec planes we flew in to Wemindji bear a goose logo. The stories of the people and the geese are thoroughly interwoven, as the Cree and the geese have continually adapted to each other and to the land. The Cree have modified certain wetlands in the rising land to attract and support geese during the spring hunt. The geese are an integral part of the story of change here, as they, like the people, have responded to the advent of guns, bush planes, roads, and hydropower development. Through it all, the Canada goose has remained at the very core of Cree culture.

I ask Dorothy if she might translate a small bit of Leopold's prose. It is not from "The Land Ethic" but from an essay called "The Geese Return." Dorothy effortlessly translates into rhythmic Cree text she had never even seen before. Leopold's theme was connection: "By [the] international commerce of geese, the waste corn of Illinois is carried through the clouds of the Arctic tundras, there to combine with the waste sunlight of a nightless June to grow goslings for all the lands in between. And in this unusual barter of food for light, and winter warmth for summer solitude, the whole continent receives as net profit a wild poem dropped from the murky skies upon the muds of March."[30] After she completes the passage, Dorothy turns to me and quietly remarks, "That's so beautiful."

Through all the changes in our lives, our times, and our places, amid all the dimensions that we must bear in mind as we chart our way forward together, there are those things that connect us: our humanity, the creatures and our stories about them, the beauty and the injury we bear witness to, the spirit that animates us, and the land that supports us all, everywhere, all the time.

**Notes**

1. G. Whiteman, "The Impact of Economic Development in James Bay, Canada," *Organization & Environment* 17 (2004): 425–448. An alternative spelling for the term is *uuchimaau*. The common English term is *tallyman*, a reference to the role of the *ouchimaw* in counting beaver lodges for the Hudson Bay Company within his hunting territory—part of the company's historic efforts to address declines in beaver populations.
2. E. Pielou, *After the Ice Age: The Return of Life to Glaciated North America* (Chicago: University of Chicago Press, 1991). The retreat of the glaciers and the subsequent rebound of the earth continues to shape landscapes and coastlines throughout the North. For example, Lakes Superior, Michigan, and Huron were a single Great Lake until the rising earth began to part the waters two thousand years ago. Because of the differential rate of rebound in the Great Lakes basin, Superior has been gradually tipping southwestward, sloshing toward Duluth.

3. The Web address of the Paakumshumwaau-Wemindji Protected Area Project is www.wemindjiprotectedarea.org.
4. For an anthology built around the critiques, see J. Callicott and M. Nelson, eds., *The Great New Wilderness Debate* (Athens: University of Georgia Press, 1998). See also M. Chapin, "A Challenge to Conservationists," *World Watch* 17, no. 6 (2004): 17–31; P. West et al., "Parks and Peoples: The Social Impact of Protected Areas," *Annual Review of Anthropology* 35 (2006): 251–277; and K. Redford et al., "Parks as Shibboleths," *Conservation Biology* 20 (2006): 1–2. Also see F. Mulrennan and F. Berkes, "Protected Area—Policy Framework," prepared for Paakumshumwaau-Wemindji Protected Area Project (July 2006).
5. The complicated story of change and adaptation among the James Bay Cree has been explored and debated widely over the last generation. See (among others) C. Martin, *Keepers of the Game: Indian-Animal Relationships and the Fur Trade* (Berkeley: University of California Press, 1978); A. Tanner, *Bringing Home Animals: Religious Ideology and Mode of Production of the Mistassini Cree Hunters* (London: Hurst, 1979); C. Bishop, "The Western James Bay Cree: Aboriginal and Early Historic Adaptations," *Prairie Forum* 8 (1983): 147–155; R. Brightman, "Conservation and Resource Depletion: The Case of the Boreal Forest Algonquians," in *The Question of the Commons: The Culture and Ecology of Communal Resources*, ed. B. J. McCay and J. M. Acheson (Tucson: University of Arizona Press, 1987), pp. 121–141; R. Brightman, *Grateful Prey: Rock Cree Human-Animal Relationships* (Berkeley: University of California Press, 1993); S. Krech, *The Ecological Indian: Myth and History* (New York: Norton, 1999); and F. Berkes, "Indigenous Knowledge and Resource Management Systems: A Native Canadian Case Study from James Bay," in *Property Rights in a Social and Ecological Context: Case Studies and Design Applications*, ed. S. Hanna and M. Munasinghe (Washington, DC: Beijer International Institute of Ecological Economics and the World Bank, 1999), pp. 99–109.
6. J. Milloy, *"National Crime": The Canadian Government and the Residential School System, 1879 to 1986* (Winnipeg: University of Manitoba Press, 1999).
7. See C. Meine, "Conservation Movement, Historical," in *Encyclopedia of Biodiversity*, vol. 1, ed. S. Levin (San Diego, CA: Academic Press, 2001), pp. 883–896.
8. See B. Minteer, *The Landscape of Reform: Civic Pragmatism and Environmental Thought in America* (Cambridge, MA: MIT Press, 2006).

9. See C. Meine, "Leopold's Fine Line," in *Correction Lines: Essays on Land, Leopold, and Conservation* (Washington, DC: Island Press, 2004), pp. 89–116.
10. A. Leopold, *Game Management* (New York: Charles Scribner's Sons, 1933), p. 422.
11. A. Leopold, "Coon Valley: An Adventure in Cooperative Conservation," in *The River of the Mother of God and Other Essays by Aldo Leopold*, ed. S. L. Flader and J. B. Callicott (Madison: University of Wisconsin Press, 1991), p. 218.
12. A. Leopold, "Engineering and Conservation," in *River of the Mother of God*, p. 254. The original address was delivered at the University of Wisconsin in 1938.
13. A. Leopold, *A Sand County Almanac and Sketches Here and There* (New York: Oxford University Press, 1949), p. viii.
14. For background on Leopold's concept of land health, see E. Freyfogle, *Bounded People, Boundless Lands: Envisioning a New Land Ethic* (Washington, DC: Island Press, 1998); A. Leopold, *For the Health of the Land: Previously Unpublished Essays and Other Writings*, ed. J. Callicott and E. Freyfogle (Washington, DC: Island Press, 1999); and J. Newton, *Aldo Leopold's Odyssey* (Washington, DC: Island Press, 2006).
15. The Web address of Chicago Wilderness is http://www.chicagowilderness.org. See R. Platt, "Chicago Wilderness: Flagship of the Urban Biodiversity Movement," remarks to the Chicago Wilderness tenth anniversary, May 17, 2006, at http://www.umass.edu/ecologicalcities/events/CWremarks.pdf. *Chicago Wilderness* magazine (http://chicagowildernessmag.org) provides continuing coverage of the consortium's work.
16. The Web address of the Quivira Coalition is http://www.quiviracoalition.org.
17. The Web addresses of these organizations are http://www.greateryellowstone.org and http://www.blackfootchallenge.org.
18. For a more complete and developed discussion of these trends in conservation, see B. Minteer and R. Manning, eds., *Reconstructing Conservation: Finding Common Ground* (Washington, DC: Island Press, 2003), especially its concluding chapter, "Finding Common Ground: Emerging Principles for a Reconstructed Conservation." See also G. Meffe et al., *Ecosystem Management: Adaptive, Community-Based Conservation* (Washington, DC: Island Press, 2002); and C. Meine, "The Once and Future Land Ethic," in *Correction Lines*, pp. 210–221.
19. C. Meine, "Conservation and the Progressive Movement," in *Correction Lines*, pp. 42–62.
20. G. Marsh, *Man and Nature; or, Physical Geography as Modified by Human Action* (Cambridge, MA: Harvard University Press, 1965; originally published 1864), p. 200.

21. In this context, see especially Leopold, "Coon Valley."
22. R. Brewer, *Conservancy: The Land Trust Movement in America* (Hanover, NH: Dartmouth University Press, 2003), p. 17.
23. Brewer, *Conservancy*, p. 32. The Nature Conservancy was founded in 1946 as the Ecologists Union and changed its name in 1950. Brewer notes, "Other well-known national land trusts were formed much later, the Trust for Public Land in 1972, the American Farmland Trust in 1980, and the Rails-to-Trails Conservancy in 1985" (p. 32).
24. G. Hardin, "The Tragedy of the Commons," *Science* 162, no. 3859 (December 13, 2009): 1243–1248.
25. See, for example, A. Leopold, "Helping Ourselves," *River of the Mother of God*, pp. 203–208. This was a significant theme in Leopold's work in the 1930s especially. He surveyed the movement in several papers, including "Farmer-Sportsman Set-ups in the North Central Region," *Proceedings of the North American Wildlife Conference*, February 3–7, 1936, Washington, DC (Washington, DC: Government Printing Office, 1936), pp. 279–85; and "Farmer-Sportsman: A Partnership for Conservation," *Transactions of the 4th North American Wildlife Conference*, February 13–15, 1939, Detroit (Washington, DC: American Wildlife Institute, 1939), pp. 145–149, 167–168. A particularly venerable example of cooperative wildlife management is Minnesota's North Heron Lake Game Producers Association, which has been meeting continuously since its founding in 1906; see http://www.nhlgpa.org.
26. See P. Annin, *The Great Lakes Water Wars* (Washington, DC: Island Press, 2006).
27. A. Leopold, "The Ecological Conscience," in *River of the Mother of God*, p. 345. See R. Knight, "Aldo Leopold, the Land Ethic, and Ecosystem Management," *Journal of Wildlife Management*, 60 (1996): 471–474; J. Callicott, "Aldo Leopold and the Foundations of Ecosystem Management," *Journal of Forestry* 98 (2000): 5–13; and Newton, *Aldo Leopold's Odyssey*.
28. Y. Shu et al., *Aa-Wiichaautuwiihkw: Coming Together to Walk Together. Creating a Culturally Appropriate Watershed and Marine Protected Area in Paakumshumwaau (Old Factory) James Bay, Quebec* (2005), p. vi. Available at http://www.wemindji protectedarea.org/assets/pdf/Aa_wiichaautuwiihkw_Report_2005.pdf.
29. M. Gnarowski, ed., *I Dream of Yesterday and Tomorrow: A Celebration of the James Bay Cree* (Kemptville, Ontario: Golden Dog Press and the Grand Council of the Cree, 2002), pp. 11–12. See also F. Berkes, "Environmental Philosophy of the Cree

People of James Bay," *Traditional Knowledge and Renewable Resource Management in Northern Regions*, ed. M. Freeman and L. Carbyn (Edmonton, Alberta: Boreal Institute for Northern Studies, University of Alberta, 1988), pp. 7–21; and P. Brown, "A Gift from the North: An Ethic for the Ecozoic," paper presented at the meeting between the Wemindji Cree and McGill University mentioned in the text.

30. Leopold, *A Sand County Almanac*, pp. 22–23.

# PART I

## *Agencies and Institutions: The Need for Innovation*

Part I begins with a candid look at federal agencies and public lands that deal with natural resources management and then illustrates the emerging models for conserving and protecting open lands. In chapter 2 Antony S. Cheng, a social scientist and forester at Colorado State University, takes a hard look at the premier federal agency in natural resources management, the U.S. Forest Service. His message is hopeful but suggests that change and innovation are needed if the Forest Service is to once more be a legitimate player in natural resources management. Cheng believes the question is not whether the agency wishes to share its authority over the resource—that time is gone. What is needed today is for natural resource agencies to plan and manage within larger ecological, economic, and sociopolitical landscapes, with funding packages that provide incentives for collaboration and stewardship with diverse groups, agencies, and individuals.

Chapter 3 is by Daniel Kemmis, former Montana political leader and founder of the Center for the Rocky Mountain West at the University of Montana. If changes are taking place within natural resource agencies, comparable changes are occurring in how public lands are managed. Kemmis offers ideas for a new paradigm of public land management, one that is more inclusive of diverse stakeholders, more respectful of economies and cultures that have historically used the lands, and more willing to experiment and learn adaptively.

Jonathan Adams, a conservation biologist and writer with The Nature Conservancy and a leading thinker in protected-area management, points out in chapter 4 that the administrative boundaries of parks and protected areas do not encompass the ecosystem boundaries of landscapes we wish to conserve. He offers the hope that if we combine conservation science with the power of local communities, protected landscapes will be both more ecologically healthy and more socially resilient in the face of ongoing change.

Following those three chapters are four case studies by practitioners and conservationists from Zimbabwe, New Zealand, Pennsylvania, and Colorado. Each tells a story of innovation—of government, private landowners, organizations, and communities partnering to promote robust economies while stewarding land- and seascapes for the benefit of communities of place and communities of interest. Collectively, the authors' stories give promise to a new way of practicing conservation, one that is pragmatic yet inclusive, and offer hope for the pressing challenges bearing down on natural landscapes.

CHAPTER TWO

# NATURAL RESOURCE AGENCIES: THE NECESSITY FOR CHANGE

*Antony S. Cheng*

The Uncompahgre Plateau in western Colorado is a sprawling uplift encompassing about 1.4 million acres. While the U.S. Department of Agriculture's Forest Service (USFS) manages the upper reaches of the plateau, the Department of the Interior's Bureau of Land Management (BLM) manages the elevation band between 6,000 and 8,000 feet. Private lands dominate at 4,500 to 6,000 feet. Large areas of the plateau are outside the historic range of ecological conditions that persisted in the area before wildfire suppression, high-grade logging, and livestock grazing altered the vegetation of the plateau. According to the Colorado Division of Wildlife (CDOW), what was once prime habitat for mule deer is now prime elk habitat. The encroachment of invasive nonnative plants has added to the ecological mix.

Restoring the plateau to within its historic range of variability clearly requires collaboration across agencies and landowners and an enormous budget. No one entity could improve the situation on its own, given budget, staffing, and bureaucratic constraints. Enter the Public Lands Partnership (PLP). Formed in 1992 as a collaborative alternative to the burgeoning "wise use" movement, the PLP comprises local government officials; environmentalists; resource users; recreationists; and state and federal natural resource agencies such as the CDOW, the USFS, and the BLM. The PLP has been committed to collaborating to address pressing public lands issues by convening regular meetings among stakeholders, sponsoring community forums and field trips on public lands issues, and stepping in to mediate conflicts and facilitate collaborative stakeholder involvement in implementing and monitoring of public lands projects. In 2000, the PLP received a five-year grant from the Ford Foundation to further its "community-based" approach to public lands stewardship. Included in PLP's portfolio of projects was the collaborative development of a plan and set of projects to restore the Uncompahgre Plateau to within its historic range of variability while also enhancing the ability of local landowners and community members to maintain livelihoods from resource use and management on the plateau.

The PLP and its fiscal agent, Unc/Com,[1] were able to secure grants and pool federal resources in ways that federal agencies could not. Through the PLP, the USFS, BLM, CDOW, and community stakeholders were able to pool nearly $4 million for the Uncompahgre Plateau Project (http://www.upproject.org/). While that is a fraction of the funding required for a multiyear, large-landscape restoration program, it is leaps and bounds more than what would have been available for ecosystem restoration in the absence of the PLP. The PLP hired public education and collaboration coordinators to bring together agencies and community members around the table. The first large-scale landscape assessment in the region was conducted on the plateau. As of this writing, experimental vegetation treatment methods have been conducted on a paired-watershed study basis on approximately four thousand acres.

* * * * *

What has happened on the Uncompahgre Plateau is symptomatic of what is happening with public lands agencies in the United States and around the world. The scope and diversity of societal demands for land and resources have outpaced the capacity of public agencies to meet those demands. In many places, community-based nongovernmental organizations such as the PLP have stepped in not only to fill some of the capacity gaps, but also to enhance the capacity of local landowners and communities to participate in and benefit from improved public lands stewardship. Barring some massive increase in funding for public agencies, we can expect more of these types of collaborative, community-based approaches to public lands management.

## Organizational Ecology and the Twentieth-Century Model of Public Lands Management

The emerging field of organizational ecology provides us with a way of understanding what is happening to the organizational system surrounding public lands management.[2] Similar to ecosystems populated by different organisms that fill specific niches, social systems are populated by different types of organizations that fulfill certain societal needs (e.g., education, health care, national security, public lands management). The theory holds that organizations that survive are *legitimate* and *accountable*. An organization is legitimate when there is congruence between its values and actions and the values and expectations of the social system. An organization is accountable when its actions are and continue to be aligned with the values and expectations of the social system.

Long-term change in the diversity of organizations occurs through competition and selection. Organizations that do not meet the expectations of the broader social system are eventually replaced by new organizations better suited to meet societal demands. Through the lens of organizational ecology, it is possible to understand the changes public lands agencies are experiencing today and can expect in the future, specifically the growing presence and influence of local governmental and nongovernmental entities.

In the United States, for the better part of the twentieth century, there was a reliance on public agencies equipped with authorities, rules, and regulations to meet society's specific needs and demands for public land management. The legitimacy of those agencies, such as the USFS, BLM, National Park Service, and U.S. Fish and Wildlife Service, was guaranteed through legalistic mechanisms of accountability. That model has been successful in many respects. Granted, public lands agencies have been at odds with social values and expectations on some occasions, such as the clear-cutting of old-growth forests in the Pacific Northwest. At the same time, the reforestation of the eastern national forests is one of the nation's great ecological restoration stories. Watersheds, rangelands, and wildlife were, for the most part, protected from the unregulated exploitation in the early twentieth century. Diverse recreation opportunities have been made available that would not have existed without public lands agencies.

### Growing Gaps between Public Lands Agencies and Rising Societal Demands

In the last twenty years, a confluence of forces has led to changes in the organizational ecologies of public land management. Those forces include changing social values and demands that emphasize environmental protection over commodity extraction; changing values, attitudes, and demographics of agency employees; problems relating to agency culture and resistance to change; and complex legal and political constraints on public lands agencies' ability to take action. This chapter will focus on one particular force: the steady divestment from public lands agencies.

Budgets for public lands agencies have been generally flat since 1990 (figure 2.1). Budget constraints mean that positions cannot be refilled if someone leaves and that projects languish without sufficient resources. The jump in conservation and land management in 2001–2003 was mostly attributable to the nearly $3 billion infusion into wildfire management resulting from the National Fire Plan, as well as some modest increases to the National Park Service budget to deal with its maintenance backlog and to the BLM budget to facilitate expansion of oil and gas exploration.

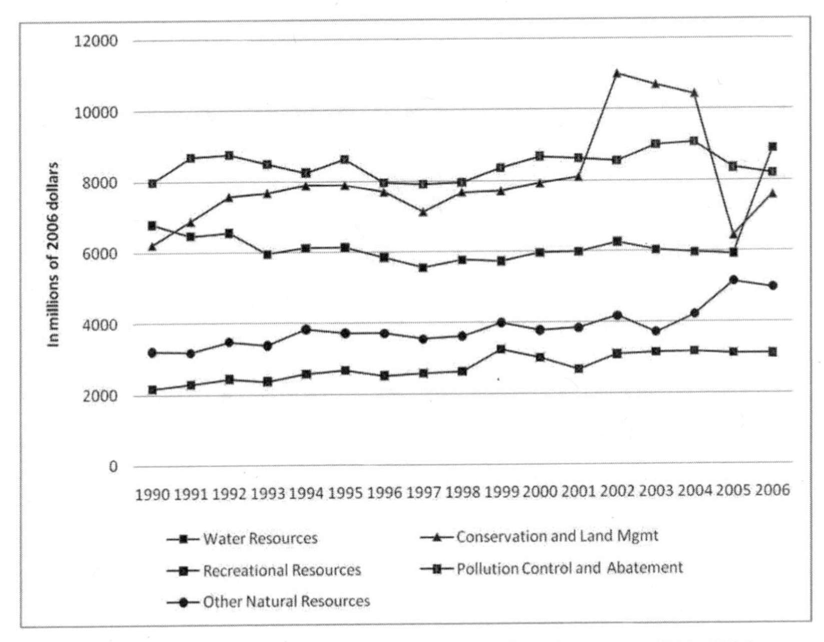


Figure 2.1 Federal investment in natural resources and environment, 1990–2006. *Source: Historical Tables: Budget of the United States Government, Fiscal Year 2008* (Washington, DC: Office of Management and Budget, 2007).

A closer look at the USFS budgets since 1991 paints a clearer picture at the agency level. Budgets for all USFS program areas except wildfire management have declined pretty steadily (figure 2.2). To make matters worse, during active wildfire seasons, the USFS wildfire suppression program "borrows" funds from all other program areas. The borrowed funds are often not returned to the original programs, causing further financial stress on already underfunded programs.[3]

The gradual divestment from public lands agencies can be seen in the declines in agency staffing. Again, the USFS provides a clear illustration. According to the U.S. Office of Personnel Management, the USFS had over 40,000 full-time-equivalent (FTE) employees in 2000. Just six years later, the number had dropped 25 percent, to just over 30,000 FTEs. The decline of their funding and staffing has limited the agencies' capacity to fulfill

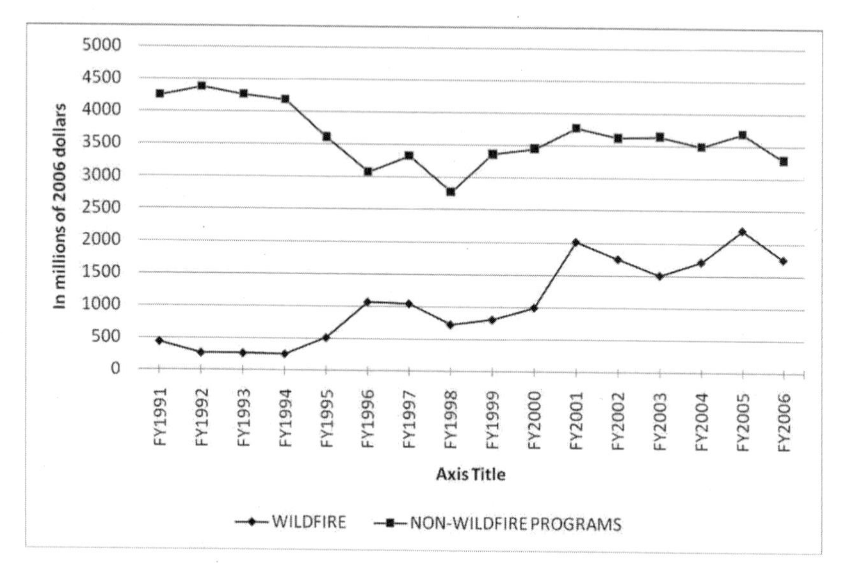


Figure 2.2 USDA Forest Service total budget vs. wildfire management budget, FY 1991–2006. *Source:* USDA Forest Service.

increasing societal expectations and demands. Given the recent fifteen-year track record, no one is foreseeing funding increases for public lands agencies.

Public lands agencies' capacity is affected not only by budgets. The Federal Activities Inventory Reform Act (FAIR) of 1998 requires federal agencies to inventory all "not inherently governmental" activities that could be competitively outsourced to private entities. The USFS competitive sourcing plan for 2005–2009 identified 21,310 FTEs—nearly two-thirds of its workforce—that could be outsourced to private entities; the jobs included everything from information collection and analysis for environmental impact studies to implementing stewardship projects for forest, rangeland, wildlife, recreation, and watershed management. These are unprecedented times for public lands agencies.

**Filling the Gaps: The Rise of Local Nongovernmental Organizations**

This is admittedly an incomplete picture that provides only a bird's-eye view of the complex functioning of public lands agencies. But the basic premise is this: Societal demands on public lands are outpacing the capacity of public lands agencies to address those demands. Evidence of this on the ground includes closing public campgrounds and imposing recreation user fees on heavily used sites. At a broader level, public lands agencies are frequently unable to address the rapid demographic, economic, and land-use changes of many communities adjacent to public land, including the following:

- The backlog of basic resource management needs, such as vegetation management, wildlife habitat improvement, watershed protection, and road maintenance or rehabilitation
- The expansion of invasive plant and animal species into native habitats
- The increased wildfire risk to lives and property
- The out-migration and attrition of people who used to work on the land and the in-migration of people who seek amenity values
- The replacement of family-wage natural resource–based jobs with lower-paying service jobs
- The increased subdivision of large private land tracts adjacent to public lands into smaller "ranchettes," leading to a decrease of affordable land and housing
- Increases in unmitigated recreation impacts, especially off-highway vehicle uses

Public lands agencies' capacities and societal demands on public lands will continue to go in opposite directions in the foreseeable future. As the capacity-demand gap widens, public lands agencies face crises of legitimacy, even as they struggle to remain accountable to the many environmental compliance policies and procedures. Following the organizational ecology perspective, the decline of one organization creates a vacuum, and other organizations will fill the niche. This is not to imply that public lands agencies are going away, but they will assume very different organizational forms than they currently have.

New organizational forms are already filling capacity-demand gaps left by public lands agencies. They are primarily community—or region—based and emphasize collaboration and partnerships, as the PLP does. They include nongovernmental organizations (NGOs) as well as state and local governments. The NGOs range from grassroots, community-based organizations to large national organizations such as The Nature Conservancy. They include both for-profit and not-for-profit organizations, tend to be community based rather than nationally based, and are private rather than public.

These new organizations and collaborative arrangements are not simply replacing what public lands agencies used to do or are expected to do. Many are tying together public lands management with broader social and economic goals, especially in communities and on private lands adjacent to public lands. In this way, these organizations are *broadening the concept of legitimacy* to encompass community, social, and economic sustainability and *redirecting accountability* away from formal legal and policy institutions to local communities and people who live, work, and play in and around public lands.

These efforts are multidimensional and can be summarized into four themes:

- Facilitating *collaboration and coordination* among governments, NGOs, and individuals to prioritize land management needs and integrate local participation
- *Pooling resources* for assessment, design, implementation, and monitoring of projects
- *Providing services* (e.g., landscape assessments, ecological and socioeconomic monitoring, public education, data collection and analysis) in the absence of available agency resources or capacity
- Assuming financial risk for conservation *entrepreneurship development* and assisting in *business training and marketing*

Community-based, collaborative approaches facilitated by NGOs are one of the fastest-growing sectors in conservation in the United States, the growth appearing to be inversely proportional to the decline in agency capacities. A preliminary search of community-based conservation NGOs in the western United States yielded over eleven thousand organizations, a majority of which were created since 1990.[4] While the number of organizations focusing on integrating ecological, economic, and social concerns is likely much smaller, it is certainly in the hundreds and growing every year.

Some might lament this diffusion of legitimacy and accountability away from public agencies to a host of local governmental and nongovernmental organizations. Public lands, after all, belong to all citizens, and the only rational—perhaps even constitutional—organizational form to legitimately manage those lands with public accountability is that of public agencies. On the other hand, if social values and expectations are not met by public agencies but can be met by new organizational forms, it is worth examining the possibilities. The reality of public lands management requires society to be open to experimentation.

## Adapting and Enduring Public Lands Agencies

Public agencies could reestablish their legitimacy by taking one of two approaches: (1) adapt their values and actions to correspond with societal values and expectations, or (2) change societal values and expectations. Their best chance is the former, but agencies would still struggle with how to address the community and regional social and economic issues tied to public lands even if they received all the funding they wished for. Those issues are simply not their forte. There are few position descriptions in public lands agencies that specifically target community-based, collaborative program development and implementation. However, if public lands agencies are to adapt and endure, some changes are necessary. Below are some general concepts condensed from ongoing public lands policy and management reform efforts.[5]

### *A National Learning Initiative*

With some NGOs and collaborative partnerships having nearly fifteen years of experience, there is no need to reinvent the wheel. While comprehensive conferences have showcased such efforts,[6] none of them have been critical, synthetic, and focused on organizational adaptation through learning. A national learning commission for collaborative conservation could be established to continuously learn from the field and revise the role of public lands agencies. There are numerous cases to draw from, such as the resource advisory councils, coordinated resource management plan groups, the Collaborative Forest Restoration Program in New Mexico, the Valles Caldera experiment, hundreds of ongoing community wildfire protection plan efforts, stewardship contracting collaborative projects, and a host of examples in multiparty assessment and monitoring of ecological and socioeconomic conditions. A national learning commission could help agencies identify best practices and productive cooperative agreements, coordinate resource management arrangements, and collaborative partnerships in which local individuals and organizations could contribute resources, expertise, and political and social support for public lands projects.

### *Spanning Boundaries*

Public lands agencies have historically been one-stop shopping centers for public lands management. Almost everything was done in house, from resource assessments and workforce training to project design and public outreach. Today and into the future, there is an ever-present need to cross jurisdictional and land ownership boundaries and take a regional approach to address the link between public lands and community and regional socioeconomic well-being. Many regional organizations, such as councils of governments and resource and conservation development councils, are spanning these boundaries and providing public agencies with pathways for integrating public lands planning and management into broader social, economic, and ecological contexts.

Another set of boundaries to cross is boundaries that separate disciplines, especially between natural and social sciences. Much of the expertise needed to deal with public lands issues involves the social sciences—community and regional economics, community sociology, interpersonal communications, group organization and management, conflict management, and the like. Spanning the natural science–social science divide requires public lands agencies to invest the human and financial resources to correspond to collaborative and cooperative initiatives—creating positions and hiring for social science expertise, especially at the field level.

Last, public lands agencies need to break down the perceived boundary defining who is an "expert" and who is a "layperson." In public lands management, the technical sciences (e.g., biology, hydrology, engineering) are valued over indigenous, experiential knowledge and other ways of knowing. This preference creates social divisions and conflict between agencies and adjacent landowners and community residents. Public agencies can minimize the divisions by facilitating opportunities for social learning and dialogue, especially in the context of landscape assessments and multiparty monitoring.

*Agency Incentives for Collaboration*

Agency resource managers are evaluated and rewarded based on performance measures. One of the key measures is quantitative outputs, such as acres treated for wildfire threats. One measure that is lacking is anything relating to building and sustaining collaborative partnerships, even though those partnerships are increasingly serving as agencies' lifelines to achieving quantitative outputs. Measures such as number of projects developed through multiparty collaborative processes and amount of financial resources pooled through collaborative partnerships could be used to "incentivize" collaboration within public lands agencies. Another way to ensure that agency managers emphasize and are rewarded for collaboration in their everyday work is to connect project funding to the presence of a collaborative group or process to design, implement, and monitor the project.

*Building Capacity of Local Individuals and Organizations*

Fortunately, public lands agencies are not at square one when working with communities, landowners, and other stakeholders in public lands management. Most agencies have histories of resource advisory committees, cooperating agency agreements, and other types of working relationships with nongovernmental entities. Moreover, landowners, community residents, and public lands stakeholders have become much more sophisticated and knowledgeable about public lands and resource issues. Harnessing that existing capacity (rather than resisting it) is not only prudent but may be necessary for the future of public lands agencies.

One loss that has been felt especially by communities working in the context of national forest lands is the elimination of the Economic Action Program (EAP) within the Cooperative Forestry program of the USFS. The EAP was never large; its budget averaged about $30 million per year. However, it provided seed money for many of the most enduring collaborative efforts and community-based NGOs, such as the PLP. With the elimination of the EAP, there is no source of financial resources to stimulate the establishment of community-based NGOs that can work side by side with public lands agencies. There is now a greater need than ever for an EAP-type program—whatever its form—for building local capacity for stewardship. This is not something the agencies can resurrect on their own—the White House and Congress must get in on the act.

Public lands agencies can go one step further than harnessing local capacity by catalyzing local government and nongovernmental sector leadership. Regular interactions between federal agencies, state agencies, local governments, and stakeholder groups send clear signals that public lands agencies are ready and willing to work with local leadership and organizations to link public lands and community economic and social health. One of the long-term, tangible benefits of catalyzing local leadership is that it can cultivate champions and constituents for public lands agencies.

*Fostering a Social Ethic Built on Reciprocity*

Public lands management requires a social ethic as much as a land ethic. The social ethic is grounded in the need to create and foster reciprocal relationships between those who manage public lands and those who use, benefit from, and enjoy them. For the agencies, it can begin with public involvement. The traditional approach to public involvement has been to invite interested and affected publics to provide input to an agency proposal during a formal "scoping" process and then to allow the public to review and comment on the draft decision. Several weeks or months could expire between the scoping and review-comment periods, with the public often left in the dark during the interval. When the final decision was released, the public might not recognize where its input and comments made a difference. In many instances, the decision would be appealed or even litigated.

That public involvement approach can be seen as extractive, with interested and affected individuals and groups expending time and energy to provide input but with little or no tangible returns besides a large amount of paperwork. Agency staffs also expend an enormous amount of time and energy with few tangible rewards. When a situation allows both agency staff and the public to leave gratified is when immediate results will be seen from their collective work. The recommendation here is simple: Devote a small amount of personnel and financial resources to small projects that yield immediately attainable results. Such projects could include a small acreage land management treatment or a multiparty monitoring project. While these small projects may be relatively insignificant at a landscape scale, they could produce substantial social results, such as increased levels of trust, the holy grail of agency public relations.

Abundant evidence from social science research bears this out; reciprocity breeds trust, which, in turn, produces even greater levels of reciprocal behavior. Appeals and litigation will always be tools, but over the long term, when agencies reciprocate the efforts of interested and affected publics, they establish pathways for increased levels of trust. Public trust in public agencies is the foundation of legitimacy and accountability.

## Conclusion

Public lands agencies have many options available for remaining legitimate and accountable. However, Congress and the White House also bear responsibility for making those options reality. Congress has already created authorizing legislation to delegate goal and priority setting to local levels, such as the provision for community wildfire protection planning under the Healthy Forest Restoration Act, the Community Forest Restoration Act, and stewardship contracting authorities. What is lacking is a clarification of the role and opportunities for national public involvement—after all, the lands in question are public lands, not just local community lands.

There is also room to locate public lands planning and management within larger regional, ecological, economic, and sociopolitical contexts. Special funding packages and incentive programs can be created for agency units that work collaboratively with regional and local entities to develop and implement cross-boundary land management plans and projects. Some creative budget restructuring would need to take place to provide adequate funding for collaborative planning, implementation, and monitoring. The next ten years will be challenging ones for public lands agencies, with continued constraints on budget and staffing resources. At the same time, if the current trends in the agencies' capacities continue, we will see more community-based and regional NGOs emerge and assume a greater role in public lands stewardship.

**Notes**

1. The PLP is not an incorporated organization; to receive and distribute funding, it created Unc/Com as a separate entity with 501(c)(3) nonprofit status under the *Internal Revenue Code.*
2. Organizational ecology is a relatively new subfield of organizational theory. As such it has both insights and limitations. My intent is to use its general concepts as an interpretive framework for the transformations occurring in public lands management.
3. GAO, *Wildfire Suppression: Funding Transfers Cause Project Cancellations and Delays, Strained Relationships, and Management Disruptions*, report no. GAO-04-412 (Washington, DC: General Accounting Office, 2004).
4. The search used GuideStar, a comprehensive database of all organization with nonprofit 501(c)(3) status under the *Internal Revenue Code.* The searches were confined to organizations with NTEE (National Taxonomy of Exempt Entities) codes C30 to C36.
5. For example, the Rural Voices for Conservation Coalition (http://www.sustainablenorthwest.org/rvcc/) brings together diverse organizations throughout the western United States to lobby Congress and executive agencies to bring public lands agencies closer to community-based conservation approaches.
6. For example, the Joint Ventures: Partners in Stewardship conference, November 17–20, 2003, in Los Angeles, and the White House Conference on Cooperative Conservation, August 29–31, 2005, in St. Louis.

CHAPTER THREE

# PUBLIC LANDS: BETTER POLICIES FROM BETTER POLITICS

*Daniel Kemmis*

In the summer of 2005, Howard Dean, the newly elected chair of the Democratic National Committee (DNC), attended a meeting of the DNC's Western Caucus in Helena, Montana. Dean's visit was meant to encourage western Democrats to participate in a partywide effort then under way to identify a handful of key themes that could serve as the core of the national party's message through 2006 and into 2008. There had to be such a lowest-common-denominator message, of course, if a Democratic presidential candidate was to have any chance of winning in November of 2008. But it was already apparent during that summer of 2005 that it would be challenging to find a message that would work in Montana and Nevada . . . and in Michigan and Pennsylvania.

At one point in the afternoon, one of Dean's aides told me, "We just need to keep anything out of our platform that would upset westerners on subjects like public lands." I wondered if it might not be possible to go a step further and actually include something that would recognize and support the good, creative work so many westerners were doing in their communities and watersheds, but the occasion didn't seem to lend itself to such a suggestion.

The Montana Democrats threw a party for Dean that evening, inviting him to a lakeside barbecue on the shoulder of the Rockies. After a long day of politics, everyone showed up thirsty, and the ice-filled tubs of beer soon attracted a crowd. But it seemed to me that people were taking longer to find the right beer than I would have expected. When I finally reached the tub, I saw why. There were Budweisers and Miller Lites . . . and that was all. Nothing wrong with offering those beers, and the populist gesture was both obvious and, in its way, good politics. But where were the local mountain microbrews—the Trout Slayer and the Moose Drool?

* * * * *

The beer tub reminded me of my earlier conversation with Dean's aide. As a Democrat who has perpetrated his share of partisan politics through the decades, I wanted to see Democrats continue to make gains throughout the Rocky Mountain West, and I thought they were more likely to do that if they could mix some homegrown positions on western issues in among the necessarily broader, more generic, nationally resonant positions that Dean was trying to identify. And at a level even deeper than my partisan loyalty, I thought the West should be trying to find a way to inject its issues into the national debate if those issues were ever to be addressed in a new and more constructive way.[1]

The good news is that westerners have been generating those new and more constructive ways of dealing with their issues for some time now, and this book is full of examples of that generative work in one crucial arena: land and resources. If it is true that a new conservation paradigm is falling into place across America, a key component of that paradigm is precisely the recognition that the way people care for their places is shaped by and in many ways specific to those places. The question is how far a new paradigm of place-based conservation can proceed without the support of a more intentional place-focused politics.

**Why Politics Matter**

Place-based conservationists tend to be so focused on solving problems on the ground that they turn their attention to politics only when they absolutely must. Not surprisingly, their political engagement, when it occurs, is more likely to be with local than with national or even state politicians. Frustrated by the failure of national politics to pay much more than lip service to collaborative approaches to land and resource issues—while perpetuating endlessly the old, tired, and tiresome ideological polarities—the new breed of place-based conservationists has fallen naturally into a strategy of ignoring Washington, D.C., whenever possible and concentrating on the hard-won but more promising working relationships with smaller jurisdictions.

That strategy, though, not infrequently runs afoul of the facts that much of the land in question is federally owned and that the policies by which it is managed are set in Washington, D.C. Occasionally, the local collaborators may go to Washington for approval of a place-based solution, as the Quincy Library Group did in 1996 or as the collaborative group working in the Black Hills did in 2002. But for understandable reasons, Congress has become increasingly wary of such case-by-case approvals of collaborative plans. That is why in the course of learning, by trial and error and sheer stubbornness, how to work together more effectively on the problems they face in their communities, their watersheds, their ecosystems, conservationlists have also generated some innovative and promising suggestions about how to make more room for these place-based, collaborative approaches in the hidebound and often dysfunctional bureaucratic structure.

The bureaucratic dysfunctions and the corresponding need to examine new approaches to public land management have been gaining broader recognition from both sides of the political aisle for some time. Former secretary of the interior Cecil Andrus has described the public land governance, a Democrat, system as "the tangled web of overlapping and often contradictory laws and regulations under which our federal public lands are managed."[2]

Republican Congressman Scott McInnis, former chair of the Subcommittee on Forests and Forest Health, describes "a decision-making apparatus that is on the verge of collapsing under its own weight."[3] Former Forest Service chief Dale Bosworth spoke of this phenomenon as "analysis paralysis."[4]

The Forest Service itself recognizes the all but impossible circumstances under which it operates. In 2002 Chief Bosworth presented to Congress a report called *The Process Predicament*, which described the effects of regulatory and administrative gridlock on national forest management.[5]

The report focused heavily on the agency's increasing inability to fulfill its primary duties. With regard to the agency's obligation to maintain forest health, the report stated, "Large portions of the National Forest System are in poor or declining health" (p. 42) and concluded that "the Forest Service operates within a statutory, regulatory, and administrative framework that has kept the agency from effectively addressing rapid declines in forest health" (p. 7).

Throughout the report, the Forest Service maintained that one result of the "process predicament" is that the health of forests is suffering. That link may be debatable, but it is hard to argue with the fact that the health of civic culture and public discourse is seriously undermined by the time-consuming and often frustrating procedures that are now typical of daily life in and around the Forest Service.

The effect on morale within the agency is no less pronounced. People cannot be expected to remain enthusiastic about their work when its results often disappear into a procedural quagmire. Finally, while several factors—including global market forces—have contributed to the decline in the timber economy in many communities surrounded by public land, many local residents are frustrated over the paralysis they see within the Forest Service.

## Policy Portfolio

Given the amount of complexity that has been built into the public lands system over the years, a "diversified policy portfolio" may be the best path forward. Investors dealing with similar complexity keep some money in stocks, some in bonds, and some in real estate to maximize the chances of substantial gains while diminishing the risk of losing all their investments. By a similar logic, a public lands policy portfolio should probably now include at least three elements: (1) comprehensive review of the entire public lands system, (2) incremental reform of the system, and (3) a deliberate period of experimentation.

The first element might involve creating a twenty-first-century form of a public land law review commission. It has been nearly forty years since the last public land law review was commissioned by Congress, making this the longest period that the system has gone without comprehensive review.

Beginning a new public land law review raises a number of questions, including how it would be formed, who would be included, and what the political costs would be. If it were proposed and supported by strong, credible, and, above all, bipartisan congressional leadership, a comprehensive review of laws might get to the heart of many problems facing the public land agencies. It could at least attempt to systemically address the basic structural problems that plague them.

While considering large-scale solutions to the operating system, attention should also be paid to opportunities for change on a more immediate scale. An example of this type of incremental reform was the bipartisan congressional effort to address the problems associated with the failing Payment in Lieu of Taxes (PILT) program.

Because people on both sides of the political fence saw the need to address the problems of an incentive structure that encouraged local governments to push for unsustainable levels of timber harvest, while still leaving many of those governments with dwindling revenues from the public lands, the political capacity developed to successfully ammend the program. The result was the bipartisan Secure Rural Schools and Community Self-Determination Act of 2000, which gave local communities incentives to balance restoration work with more sustainable levels of harvest while stabilizing their revenue streams. Congress, agencies, and other interested parties should continue to look for similar opportunities to address acute problems and bring positive incremental change to the system.

The third element of the proposed policy portfolio, deliberate experimentation, is reflected in several recent proposals to try new approaches to managing public lands. Many of the proposals call for legislatively authorized experiments or pilot projects that are to be implemented, monitored, and evaluated through various forms of collaborative governance.

**New Models of Public Land Management**

Three examples will illustrate these new governance and management ideas: the Forest Options Group's Second Century Report, the "Mugido" proposal of the Quivira Coalition, and the Region Seven proposal that emerged from the Lubrecht Conversations.

One of the most rigorous suggestions for innovation came from the Forest Options Group, a consortium of environmentalists and timber industry representatives. The group convened in the late 1990s to discuss and address the growing concern about the viability of Forest Service operations. Finding that the existing budgetary and administrative framework of the Forest Service encouraged "polarization, not cooperation," the group suggested that alternative budgeting mechanisms and governance structures—to reduce both subsidies and inefficiencies—might restore public trust in the agency. Arguing that identifying the best combination of management and budget structures was an "empirical question," the group developed five pilot models and recommended that each model be tested on two national forests. Four of the models focused on incorporating a revised budgeting plan with no changes in management, while the fifth presented a "collaborative governance" option, in which the forest plan would be written and the forest supervisor hired by a local board. The managers of the forests would adhere to all environmental laws but would be given discretion in whether to follow internal agency directives. All the pilots would be evaluated for effectiveness after five years.

Meanwhile, the so-called Mugido proposal for new management structures emerged from the Quivira Coalition. The coalition's executive director, Courtney White, explains, "A mugido is a stretch of public land where the government vastly reduces its regulatory role in exchange for high environmental stewardship by a private entity. In a mugido the government's role is to set ecological and social standards and objectives through collaborative goal-setting, provide technical assistance (fire, archaeology, biology), and conduct oversight and monitoring. The role of the private entity is to meet, or exceed, the collaboratively-derived goals and objectives."[6] The land in question would remain part of the public domain, where it would still be owned by the American public and managed to attain national goals, while being operated by a private entity in collaboration with the federal land agency.

"For example," White explains, "while the Forest Service or the BLM would set environmental goals for the allotment (or landscape), it would be the permittee's decision on how to achieve them. The goals would be set collaboratively, drawing on each member's strengths, but the permittee would have discretion over the toolbox: what type of livestock to use, for example, their numbers, timing, and intensity."[7]

Permittees would be given the opportunity to be innovative, flexible, and entrepreneurial in making day-to-day management decisions but with the understanding that the government would monitor and evaluate the results and respond accordingly. "Not all regulation would disappear," White explains. "Ensuring the recovery and maintenance of endangered plants and animals, for instance, would be subject to enforcement. But collaborative decision-making coupled with innovative implementation of best management practices, audited by the government, means that the 'hammer' of regulation could be laid down."[8]

The third proposal to facilitate experimentation and collaboration on national forest lands emerged from a series of conversations held in the late 1990s at the University of Montana's Lubrecht Experimental Forest. Picking up on the fact that, because of past regional consolidation, the previous Region 7 had been absorbed into Region 6 and as a result there had not been a Region 7 in the national forest system since 1965, the Lubrecht group proposed the creation of a new, nongeographical region that would house experimental projects on national forest lands.[9] This new "virtual region" would test innovative approaches to forest management while focusing on collaborative governance structures and other mechanisms to overcome some of the problems that beset the current system of national forest governance. Such an experimental approach would not attempt to change the entire national forest system but would recognize problems and invite and test innovative solutions in a few carefully chosen settings. The first step would be to conduct a national competition for the selection of experimental projects to test new models of management or governance. A blue-ribbon commission would be organized to solicit proposals for alternative approaches to public land management and governance, select promising projects, and guide the implementation process. The projects selected would make up the new Region 7. To encourage the development and careful testing of alternative approaches to national forest management, enabling legislation for Region 7 would do the following:

- Establish a national competition for selecting promising projects
- Establish a broadly representative advisory committee to guide project selection and monitoring
- Emphasize the experimental, adaptive nature of projects
- Authorize and encourage projects across a range of administrative and geographic scales

- Require monitoring of both process and outcome against established baselines
- Require a cumulative record of project activities and outcomes
- Ensure broad dissemination of lessons learned

**Principles of New Models for Public Land Management**

The three alternative proposals for managing our public lands described in the preceding section could test a broad range of models. The types of experiments tested would depend largely on what public land stakeholders are currently attempting or would like the opportunity to try in their own communities. The following list is meant not to be prescriptive but merely to suggest the possible range and variety of models:

- *Trust model.* The public land in question would be managed by a board of trustees, pursuant to a binding trust instrument.
- *Budgetary incentives.* After some initial period of federal budgetary support, the experimental area would be expected to generate most or all of its own funds.
- *Collaborative planning model.* A collaborative body would write a management plan for the area, and existing public land managers would be charged with implementing it.
- *Collaborative governance model.* A collaborative group would be empowered to write and oversee implementation of the management plan.

If these experiments in public land management are to be of consequence, it is important that they be carefully monitored and evaluated and that the results be analyzed honestly. The resulting information must be broadly disseminated so that as many people as possible can learn from the experiments and adapt the lessons to their own settings. Any experiment should also be allowed to operate for at least five years—preferably ten or even fifteen years—so that its long-term viability can be legitimately evaluated.

## Functional Public Lands, Functional Politics

The Forest Options, Region 7, and Mugido proposals all call for the creation of space within the established institutional framework for some carefully chosen, closely monitored experiments with management and governance methods that rely more heavily on local knowledge and collaborative problem solving than on one-size-fits-all national approaches.

The other feature that all of these proposals share is that none has garnered enough support in Congress or in any national administration, Democratic or Republican, to come within striking distance of being implemented. The time has come to at least ask whether promising policy initiatives don't have to be linked to some new forms of politics. What these proposed experimental frameworks represent is an effort to make room for some place-based microbrews among the national brews in the management and governance tub. But the record seems to indicate that we won't succeed in producing that richer mix of management alternatives unless we are simultaneously being creative about getting the microbrews into the political tub. Again, the good news is that this is starting to happen.

As it turns out, John Hickenlooper, the mayor of Denver who worked so hard to bring the 2008 Democratic National Convention to Denver and who eventually welcomed the delegates to that convention, gained his initial fame in Denver through the popularity of his brewpub. The Wynkoop Brewery was one of the earliest new businesses to launch the resurrection of Lower Downtown Denver, a neighborhood that convention delegates from all around America explored between sessions of the convention. Hickenlooper's hope had been that delegates would recognize where they were—that is, not only in Denver, but also in the West. He had pitched Denver as the best convention site because he (and many other western

Democrats) thought it would be refreshing for delegates to get a sense of some of the new political and policy breezes blowing in the West.

In a speech to Montana Democrats, "Hickenlooper said his goal is to make the Democratic National Convention a regional event, instead of just focusing on Denver, and have the national media see the West features different leadership approaches. That's why Democrats are making such gains in the Rocky Mountain states, he said."[10]

Hickenlooper was talking partisan politics, which was not surprising, given that it was a political party's convention that was coming to town. But again, there was (and is) something deeper at work here than partisan politics. What the mayor was underscoring was a politics of place—a politics that recognizes and celebrates the fact that people who live in different places approach things in different ways.

Tip O'Neill's famous dictum that "all politics is local" might lull us into thinking that place-based politics is the norm. Without denying the deep and useful wisdom of that adage, it won't get us very far into this topic unless we put it up against the fact that the history of conservation in America has been distinctly and aggressively national and very often suspicious of (if not hostile to) any genuinely local politics touching on the field of conservation.

It may be helpful to remember that the field of conservation has never been—and could never have been—free of politics and that the big advances have always been intensely political, involving major new departures in politics that strongly reflected some of the major dimensions of the emerging conservation paradigms.

This is a long and complex story, which I've dealt with in much greater detail elsewhere than space will allow me to do here.[11] In a nutshell, both the conservation movement of the late nineteenth and early twentieth centuries and the environmental movement of the late twentieth century relied heavily on national politics and national policies to advance their cause. Theodore

Roosevelt, the most visible and vocal proponent of the first conservation movement (and a Republican), eventually endorsed a political philosophy that he called the "New Nationalism," which aimed to mobilize the political will to undertake public initiatives, not just at the local or state level, but as a nation. That national approach eventually became such an accepted part of American life that it has become difficult for us to understand what a major departure it was in Roosevelt's time. Nowhere did the New Nationalism leave a more lasting legacy than in the creation of the national forest system. Since that national policy initiative has survived to this day, while the politics that produced it have long since vanished, we easily lose sight of the fact that, to enact such an innovative departure in policy, Roosevelt and others first had to fashion a newly effective national politics.

## Conclusion

The minute we start trying to think beyond the national approach to conservation politics, we run into a variety of challenges and not a few box canyons. There are no road maps for the kind of place-based politics that would have to accompany place-based conservation. Still, there are glimmers, and now sometimes even shafts of light, penetrating the mist. If we concentrate momentarily on the public lands of the West, for example, we find some intriguing signs of the emergence of a more place-based politics. A number of Rocky Mountain states have now made a serious effort to align their presidential primaries and caucuses, in the hope that a "western primary" would induce presidential candidates to pay more attention to the region's most pressing issues, take constructive positions on them, and follow through if elected. If in fact there are consistent elements of an emerging conservation paradigm, and if that new form of conservation has implications for national policy, a western primary would be one important forum for examining those implications and presenting them to candidates in a way that enables them to support appropriate policy initiatives.

Without such new, place-focused political forms, the new conservation paradigm is likely to fall short of its highest promise.

**Notes**

1. More detailed thoughts and references are provided in D. Kemmis, *This Sovereign Land: A New Vision for Governing the West* (Washington, DC: Island Press, 2001).
2. Andrus Center for Public Policy, Policy after Politics: How Should the Next Administration Approach Public Land Management in the Western States? p. 2 (June 1, 2000), at www.andruscenter.org/AndrusCenter.data/Components/PDF%20FILES/PAP_whitepaper.pdf.
3. Conflicting Laws and Regulations—Gridlock on the National Forests: Oversight Hearing before the Subcomm. on Forests and Forest Health of the House Comm. on Resources, 107th Cong. (2001) (statement of Rep. Scott McInnis, Chairman, Subcomm. on Forests and Forest Health), available at www.house.gov/resources/107cong/forests/2001dec04/mcinnis.htm.
4. Conflicting Laws and Regulations—Gridlock on the National Forests: Oversight Hearing before the Subcomm. on Forests and Forest Health of the House Comm. on Resources, 107th Cong. (2001) (statement of Dale Bosworth, Chief, U.S. Forest Service), available at www.house.gov/resources/107cong/forests/2001dec04/bosworth.htm.
5. USDA Forest Service, *The Process Predicament: How Statutory, Regulatory, and Administrative Factors Affect National Forest Management* (Washington, DC: USDA Forest Service, June 2002.
6. *Quivira Coalition Newsletter*, no. 4 (April 2006): 23.
7. Ibid.
8. Ibid.
9. D. Kemmis, "Re-Examining the Governing Framework of the Public Lands," *University of Colorado Law Review* 75 (2005): 1127–32.
10. C. Johnson. "Denver mayor touts western Democrats' pragmatic approach." *Helena Independent Record,* March 11, 2007.
11. See Kemmis, *This Sovereign Land.*

CHAPTER FOUR

# PARKS AND PROTECTED AREAS: CONSERVING LANDS ACROSS ADMINISTRATIVE BOUNDARIES

*Jonathan Adams*

Cienega Creek rises in Arizona's Sonoita Valley and flows north toward Tucson, between the Sky Islands of the Santa Rita and Whetstone mountains. Near its source, Cienega Creek flows through high desert grasslands, some of the best examples of native grasslands that remain in Arizona and ideal habitat for a host of native species. Three native fish species, all officially endangered, call the Cienega Creek home, and nearly two hundred species of birds have been identified nearby. Cattle, introduced here in large numbers in the 1820s, also roam these arid lands.[1] From the Civil War era until the late 1980s, much of the Sonoita Valley was private ranch land, including two huge operations, the Empire and Cienega ranches.

Just twenty miles southeast of booming Tucson's borders, Cienega Creek flows under Interstate 10, and the situation changes. The state of Arizona owns most of the land just north of the highway and leases it for cattle grazing, largely to raise money for public schools. By law, the state must generate the highest return possible from these lands; in some cases, that may mean selling to a developer who would replace the cattle with houses. Beyond these state trust lands farther to the north lie Saguaro National Park and the Rincon Wilderness, part of Coronado National Forest. So while lands to the north and south are protected, the portion of the creek along the interstate runs through areas important for conservation but also ripe for development. A piece of the puzzle is missing.

The story of how the Empire and Cienega ranches became the Las Cienegas National Conservation Area, how the missing link may finally be found, and how they came to be seen as pieces in the conservation puzzle in the first place has much to tell us about protected areas, science, community, and the need to plan for and carry out broad-scale conservation. The intersection of science and community may be the most hopeful place in modern conservation.

In the late 1980s, the Bureau of Land Management (BLM) acquired the Empire and Cienega ranches in Sonoita Valley through a series of land swaps. The BLM leased grazing land in what was called the Empire-Cienega Resource Management Area to ranchers Mac and John Donaldson, father and son, from Sonoita. The Donaldsons would become leaders of an effort to keep the ranch in operation while maintaining its scenic and ecological value.

While the Donaldsons were out ahead of most of their neighbors in terms of running an ecologically sensitive ranch, the rest of the Sonoita Valley community was not that far behind. They created several community organizations, one of which, the Sonoita Valley Planning Partnership, began working with the BLM on a management plan for the resource management area.

The idea of allowing communities living near and using public lands to have a major say in how those lands are managed has become a common approach to conservation in the developing world—with distinctly mixed results—but it remains the exception in the United States. At Las Cienegas, the community-based approach to the management plan began with a focus on the 43,000-acre resource management area and grew from there.

The community developed a landscape vision of conservation for Cienega Creek that extended far beyond the BLM land, across I-10 and all the way to the Rincon Wilderness Area. The additional land was largely state trust land and enlarged the planning area to about 140,000 acres. The community wanted the whole thing designated a national conservation area (NCA), a congressional designation meant to conserve particularly important BLM lands for their exceptional natural, recreational, cultural, wildlife, aquatic, archaeological, paleontological, historical, educational, and scientific resources.

With all the community support for Las Cienegas, the proposed NCA got bipartisan support and endorsements from environmentalists, ranchers, even real estate developers. In 1999, Congress unanimously passed a bill to create the Las Cienegas NCA, an unheard-of event. Even more remarkably, the bill passed at a time when western congressional delegations scowled at the very mention of a new protected area under federal control.

Despite the federal land ownership, this is bottom-up rather than top-down conservation. The people of Sonoita essentially created the national conservation area, and they intend to remain involved. If the BLM backtracks or fails to live up to its responsibilities, the community will respond. Far more than the rules governing the national conservation area, it is the commitment of the people living near Cienega Creek that holds promise. Communities of people committed to husbanding land for generations so that it sustains life—cultivated and wild—form the foundation of lasting conservation.

But even with the commitment of the Donaldsons and others in the community, Las Cienegas was not a complete victory. During negotiations over the bill, Congress and the Arizona State Land Department removed the area between I-10 and the wilderness area, thus creating the missing link. Congress instead mandated preparation of an assessment to explore the regional significance of that link, now called the Cienega Corridor, and offered recommendations for its management and protection. That assessment concluded that protection of the Cienega Corridor is crucial for regional biological and economic health, as well as preservation of cultural heritage and archaeological sites.

The assessment recommended a community-based collaborative management approach for the Cienega Corridor. Partly in response, an ad hoc coalition of more than forty landowners, business leaders, land managers, and environmental advocates formed the Cienega Corridor Conservation Council. That group is developing a protection plan for the corridor, giving presentations throughout the region, developing brochures, and sponsoring media events to spread the word about the corridor and its importance for conservation.[2]

In 2004, the Cultural Landscape Foundation recognized the Cienega Corridor as one of the nation's top seven endangered cultural landscapes, expanding our conception of what a protected area means. It was chosen based on the presence of archaeological sites, ghost towns, railroad camps, and current and historic ranches and transportation routes. Those cultural resources are at risk because of rapid growth and development in the adjacent urban edge of Tucson. In its designation, the Cultural Landscape Foundation recognized not only the values of the Cienega Corridor landscape, but also the value of a diverse group of citizens, land managers, scientists, and other stakeholders who formed the Cienega Corridor Conservation Council.[3]

The efforts to protect the Cienega Corridor show promise, but there are no guarantees. The fabric of the community, its social capital, must remain intact. That lesson may well hold for any broad-scale conservation efforts. Such efforts attempt to address large, complex landscapes and often contain national parks or other protected areas, like Las Cienegas, at their core. The success of those efforts, however, may boil down not to the protected areas themselves but to the ability of communities to come together over a shared vision for the land.

* * * * *

This chapter will explore an emerging alternative approach to land conservation, one that emphasizes dealing with private landowners and local communities, rather than the top-down federal emphasis that is strictly on public lands. This new approach to protecting lands goes far deeper than just getting people to participate in conservation planning meetings. It means examining our assumptions about the role of protected areas in conservation and what those areas can and cannot do to ensure the survival of wild species and wild places.

**A New Perspective**

Parks and protected areas remain the bedrock of conservation efforts worldwide, and for good reason. Parks work. In the most thorough review of the effectiveness of national parks thus far, a study of ninety-three large parks covering nearly 70,000 square miles in twenty-two tropical countries found that 83 percent of the parks were holding their borders against agricultural encroachment. These are not just the newer parks, either; the median age of the parks in the study was twenty-three years.[4]

Parks serve an even broader purpose, however. Parks are about hope. In choosing what places to set aside permanently, we make a nearly unmatched statement of collective purpose, marking out clearly what we value and proclaiming a vision of who we want to be. As Wallace Stegner wrote, "The national park idea, the best idea we ever had, was inevitable as soon as Americans learned to confront the wild continent not with fear and cupidity but with delight, wonder, and awe."[5] But too often the parks are symbols not of hope and pride, but of fear and despair, statements not about the future but about attempts to hold back the tide of change.

The question comes down to what we see as the landscapes of our future. If we are prepared to abandon the notion of stewardship and allow all manner of destructive land uses everywhere except in those few places we call parks, then the parks must be fortresses. That is a bleak vision, for those fortresses would, like all fortresses, eventually fail. Parks should instead be the foundation stones of communities rather than fortresses, beloved places set aside because they remind us of what we must protect everywhere, rather than to protect the last of something.[6]

Yet, important as they are, both symbolically and practically, parks can take us only so far in our conservation efforts. A park, perhaps emblazoned on our imagination by Yellowstone National Park, fundamentally does one thing and does it extremely well: It prevents the land within its borders from being transformed into something entirely new—a farm, a factory, a subdivision.[7]

A park does not prevent any number of other disasters, however, from the invasion of aggressive exotic pests to the construction of towns on its borders that prevent species from moving in and out, to climate change that rearranges the ecological dynamics of entire regions and Earth itself. Effective conservation at those broader scales demands that we consider what happens both inside parks and in the landscapes in which the parks are embedded. Traditional conservation skills like wildlife management and even the more recent scientific specialties like landscape ecology will not suffice by themselves. Conservation must come to grips with the ecological context and also with the emotional, value-laden, and inescapably political context of each and every conservation landscape.

The political context must include the human communities that surround parks, as in Las Cienegas, as well as the more distant communities that value parks and wildlands as refuges or simply as visions of wilderness that they may never see. Effective conservation, therefore, means dealing with the things people do, which requires working at human scales and at scales broad enough to encompass big animals and ecological processes operating over large areas and hundreds of years.[8]

Conservation has rarely been able to function at such broad scales and has generally proceeded park by park. The limitations of that approach are now painfully clear. Yellowstone, with its nearly perfectly rectilinear shape, represents just a swatch cut from the fabric of the western landscape. Despite covering more than 3,400 square miles, Yellowstone faces threats that extend into three states, across the border into Canada, and beyond.

That vast landscape, the Greater Yellowstone Ecosystem, is by one estimate the size of Virginia.[9] Greater Yellowstone now hosts all manner of land uses, creating a complex mosaic. Some pieces of the mosaic remain more or less intact, while human activity has completely transformed others. Studies have demonstrated that such a fragmented landscape poses grave threats to all manner of species, all the more because the fragmentation is usually irreversible. But the problem is as much social and cultural fragmentation as biological or ecological fragmentation.

The success or failure of conservation across an area the size of Greater Yellowstone, or similar expanses anywhere in the world, ultimately rests with the people who live there. Yet more and more frequently the people on whom conservation will ultimately depend live not in cohesive communities, but separately and apart, without bonds of kinship and shared tradition holding them together. The loss of such social capital makes it ever harder for communities that value wild species and wild places to act on those values and provide a counterweight to powerful and entrenched political and economic interests whose values lie elsewhere.[10]

Even where social capital is abundant, conservation has traditionally overlooked, intentionally or otherwise, the needs and values of the people who live near parks. Hence, a park becomes a challenge and an invitation to conflict. In almost all cases, however, the line that defines a park reflects human convenience rather than ecological necessity; the boundary will be illusory for creatures and often for humans, as well. The line remains a necessity, because in most places we now have no choice but to draw it and make a stand. But conservation does not have the troops to defend the parks if people decide not to value them. The sooner we reach the point where we no longer need to draw lines, or where we need to draw them only as a matter of administrative convenience, the more of Earth's diversity we will be able to save.

While parks have always existed in dynamic physical landscapes, whether we recognized it or not, today they also exist within dynamic political and social landscapes. Successful conservation under those circumstances will require new appreciation for where science, community, and values intersect. They intersect when communities make value judgments about what to conserve and what to develop, with science as a guide to the consequences of their choices.

Engaging communities in decisions about the future of the land around them makes such evident sense that the power of the idea led some conservationists to get a bit overzealous in the early 1990s, deciding that community-based conservation was the only hope. The promise of the approach has not been realized, largely because simultaneously protecting animals and their habitats from some kinds of human use while allowing other uses has proven far more difficult than anyone imagined. The idea will not go away, however, because the people living on the edges of protected areas will not go away; in fact, there will be more of them. In some places, they form cohesive communities, and those communities can and should be more involved in conservation. In other places, community is far more elusive, and staking the future of conservation in those places on the hope that some common ground exists among hundreds or thousands of individuals with disparate values and desires is basically just wishful thinking.

The ethical and scientific foundations of national parks and the relationship of the parks to local communities should be subject to vigorous debate. In Las Cienegas both the community and conservation may be winners, but more often the creation of a park or protected area means someone wins and someone else loses. Who wins and who loses should be the subject of careful study. Unfortunately, nasty and unproductive debate often replaces sober analysis.[11] Instead of engaging in a meaningful conversation about where and how to use or not use land, groups with vested interests in parks and protected areas have settled into what some conservation biologists call "a dialogue of the deaf."[12]

The debate should not be about parks per se, but rather about where parks will work; where they will not work; and the ecological, cultural, and political conditions that will make the difference. The answer to any question about setting aside land depends on what you are trying to protect, for how long, and where. Sometimes a small reserve will do the job just fine, so conservation planners need to consider wildlands on many scales, from the bog on the outskirts of town to a 10,000-square-mile reserve straddling the U.S.-Canada border, roomy enough for a population of grizzly bears that can thrive for centuries.[13] Rather than searching in vain for universal rules for parks, we must learn all we can about each species, each landscape, each community. As Michael Soulé, a founder of conservation biology, told me, "The answer to every question in ecology is 'It depends,' or 'I don't know.'"

While no one concerned with conservation advocates eliminating national parks, the debate frequently reaches such a fever pitch that partisans call for everyone to choose sides—for parks or against them. The outcome of such a fight would be losses on all sides. Neither conservationists nor rural communities—whether in the rainforest or the American West or on the African savanna—have much economic or political power and would be far better off making common cause in defense of both biological and cultural diversity. Efforts to preserve one will sometimes mean diminishing the other, offering another incentive for taking a broad view that includes parks, industrial lands, and everything in between.[14]

The challenge for conservation is to think big and think small—big enough to provide room for grizzly bears and elephants, but small enough that human communities can play meaningful roles. Fundamental to any such effort is knowing what you are trying to conserve, knowing why you need to put that park or that bike trail or that housing development in one place and not another, and ultimately knowing what sort of a world we wish our descendants to inherit. The last is a question of ethics, but the first two are questions of science, for which we need a far more systematic approach to conservation than we have used so far. The answers will grow from thousands of breakfast table debates, a national and international dialogue.

**Science, Community, and Conservation**

Creating a park is an essential yet desperate act, typically based on the idea that we can by force of law ensure that what happens ecologically on one side of the lines we draw will differ fundamentally from what happens on the other. A park is also, in the words of writer David Quammen, "a wildly ambivalent act of collective purpose: dreamy yet provident, selfish yet sacrificial, local yet global in significance."[15] Park boundaries rarely if ever reflect ecological reality, because ecology has until quite recently played almost no role in deciding where to put parks. While people have been setting aside land for one reason or another for four millennia, in doing so they protected prerogative or spirituality first and nature only as an afterthought. Instead, governments and individuals set aside grand or symbolic lands, like Yellowstone and the Grand Canyon, or lands that had little economic use, like the parks of the Mountain West, brimming with rocks and snow. In the United States, federal assessments of wilderness study areas focus largely on how often people use them for recreation, not how well they protect species or ecosystems.[16]

Without either a broad vision or a rigorous scientific plan, and without the political clout that comes from connecting to the values of people who live on and work the land, conservationists have generally settled for whatever land is available. That guarantees minimalism and eventual failure; we cannot save some of everything—the most fundamental kind of conservation, used by Noah—if we take as a given that all the best land will be used for something other than conservation.

Another result of this willy-nilly conservation is that national parks and other protected areas are almost without exception too small to conserve the species within them.[17] They often are in the wrong places, as well. In the United States and most other places, opportunism rather than science drove the creation of national parks and wilderness areas. Parks usually become parks for scenery and recreation, and for what they are not—not farms, not mines, not near town. Remote and unusable land costs less to protect and usually fits at least a generalized notion of what intact and untouched landscapes look like. As a result, protected areas tend to cover high altitudes or wetlands, while productive, lower-elevation lands remain outside of the reserve system The pattern repeats over and over again: The public owns mountains and ridgetops, but individuals own the rich valley bottoms. The species and communities on those private lands have no guarantee of protection. Nationwide, private landowners hold more than 90 percent of the productive, low-elevation lands. The most thorough study so far suggests that protected areas in the United States are heavily biased, the majority of different habitats occurring only rarely in reserves.[18]

The incessant cry from industry that most any conservation effort means locking up exploitable natural resources—an exaggeration at best—reflects the belief that setting aside productive land or water is a waste, an ideology embodied in water laws in the United States that explicitly define "wasted" as water left in the river for the fish. That attitude, and the political power of big industries, places intense pressure on governments to avoid politically costly confrontation by protecting areas that don't need protection. For conservation at the national or continental scale to succeed, the design of parks and other protected areas will need to build on their ecological importance rather than their economic irrelevance.

Political steadfastness regarding parks will not suffice in any event, as climate change makes clear how incomplete parks are as tools against complex global pressures, both ecological and otherwise. Even for Yellowstone, still a symbol and an inspiration more than 130 years after its creation, the future has rarely been more uncertain. As the global climate shifts, driven largely by human activities, everything in Yellowstone—its elk, wolves, grizzly bears, quaking aspen groves, and all—will also have to move. Scientists expect that species will shift ranges poleward in latitude and up in altitude, nonnative species may invade new areas, once common species may go extinct, and once rare species may become common.[19] Whether humans, including conservationists, business owners, and policy makers, as well as the people who live near Yellowstone, foster or inhibit that movement of species and natural communities may well determine the future of Earth's wild places.[20]

If our goal is to conserve the plants and animals that make Yellowstone or any other park unique, then we must consider, to the extent the science allows, where they may be a century from now. We can secure the future for human as well as natural communities if we consider the park and its surroundings and think big. The entire Greater Yellowstone Ecosystem may be a large enough canvas on which to plan for the region's wolves and bison, as well as its ranchers and its increasingly footloose population of engineers, architects, and software designers.

Given that the precise impacts of climate change in any one spot on the map remain uncertain, the key will be keeping options open for both human development and shifting natural systems. If we are careful and focus on maintaining natural and political systems that are resilient in the face of change, new associations of species and ecosystems may establish dynamic equilibria. If we act heedlessly, however, and foreclose on the future by, for example, taking today's Yellowstone boundaries as absolute and allowing all manner of land use right up to the edge, then when systems need to shift, they will have nowhere to go. Such brittle systems may break and collapse entirely, and the results could be catastrophic.[21]

## Conclusion

If we are to build and nurture ecological and social systems that are resilient rather than brittle in the face of change, we will need to change the philosophy and practice of conservation. Just such a change has been in process for a decade, transforming conservation to include both species and broad landscapes, protected areas and the places where people live and work. Conservationists and others hoped for a time in the late 1980s and early 1990s that helping people improve their economic condition would lead as if by magic to more conservation, as well. Parks could do it all: protect nature and raise the standard of living among the rural poor. That hope has gone largely unfulfilled, and grinding poverty side by side with glorious wilderness remains a cruel taunt and the inescapable fact of modern conservation.[22]

Rather than retreat behind the park walls, or abandon them altogether, conservationists will need to work across all the relevant scales—to spend more money to set aside key land, to convince governments and communities to change the way they manage land, and to restore land that has been degraded, like Florida's Everglades.[23] None of this will be cheap or easy. The United States is enormously fortunate to have large areas of wildlands as well as the economic resources to conserve them and to be able to help individuals and communities that might be affected by conservation. Hundreds of communities in the United States have passed bond measures to preserve open space, essentially taxing themselves in the interest of the environment. Even so, the worldwide conservation community often settles for table scraps, relying on meager budgets from government and the generosity of individuals. We always seem able to find the money for new roads, new dams. We can find the money for conservation if we choose to look for it.

The greatest hope lies in combining conservation science with the power of community and a reinvigorated democracy responsive to both local and national concerns. The Jeffersonian ideal that government should rest on civic dialogue at the level closest to those who will be affected by government decisions has enormous power.[24] Local decisions, however, too often build on a purely local view of ecology. Decades of pitched battles that pit environmentalists against loggers, miners, ranchers, and developers have left deep scars that will make dialogue difficult, to say the least. Until now, however, the opposing sides had no way of seeing how their landscape fit into the broader ecological puzzle, how to judge its importance relative to the larger whole.[25]

Far more than ever before, conservationists and rural communities can find common ground, and they can develop shared visions for the future that reflect not simply local interests but also global ones, visions that form the foundation for making decisions about how the land, public and private, is to be both protected and used. No one, not even the most hard-core city dweller, wants to live in a paved-over world with no wild creatures. A shared vision will include places for wilderness, for city parks, for working farms and ranches. If science informs those visions, perhaps humanity can find a way to lasting coexistence with nature.

Once you take this perspective, the size of the challenge—and the opportunity—becomes clear. We now find ourselves at the center of a promising but highly uncertain movement, one that melds a commitment to the people who husband their land with the best thinking in conservation science. The outcome of that fraught process may be the last best hope for Earth and all her creatures.

**Notes**

1. C. J. Bahre, *A Legacy of Change: Historic Human Impact on Vegetation in the Arizona Borderlands* (Tucson: University of Arizona Press, 1991).

2. See http://rinconinstitute.org/content/view/7/43/.
3. E. Brott, "The Living Desert: Our Most Valuable Lesson?" *Terrain.org* 16 (spring-summer 2005), http://www.terrain.org/columns/16/guest.htm.
4. A. Bruner et al., "Effectiveness of Parks in Protecting Tropical Biodiversity," *Science* 291 (2001): 125–128; see A. Rodrigues et al., "Effectiveness of the Global Protected Area Network in Representing Species Diversity," *Nature* 428 (2004): 640–643.
5. W. Stegner, "The Best Idea We Ever Had," in *Marking the Sparrow's Fall*, ed. P. Stegner (New York: Henry Holt and Company, 1998), p. 135.
6. J. Adams, *The Future of the Wild: Radical Conservation for a Crowded World* (Boston: Beacon Press, 2006).
7. L. Naughton-Treves et al., "The Role of Protected Areas in Conserving Biodiversity and Sustaining Local Livelihoods," *Annual Review of Environmental Resources* 30 (2005): 219–252.
8. R. Defries, "Land Use Change Around Protected Areas: Management to Balance Human Needs and Ecological Function," *Ecological Applications* 17 (2007): 1031–1038; A. Hansen and R. Defries, "Ecological Mechanisms Linking Protected Areas to Surrounding Lands," *Ecological Applications* 17 (2007): 974–988.
9. R. Noss, "A Biological Conservation Assessment for the Greater Yellowstone Ecosystem." Report to the Greater Yellowstone Coalition, July 2001.
10. M. Moore et al., "Linking Human and Ecosystem Health: The Benefits of Community Involvement in Conservation Groups," *Ecohealth* 3 (2006): 255–261; R. Putnam, *Bowling Alone: The Collapse and Revival of American Community* (New York: Simon and Schuster, 2001).
11. R. Blaustein, "Protected Areas and Equity Concerns," *Bioscience* 57 (2007): 216–221; M. Chapin, "A Challenge to Conservationists," *World Watch* 17 (2004): 17–31; M. Dowie, "Conservation Refugees," *Orion* (November-December 2005): 1–8.
12. K. Redford et al., "Parks as Shibboleths," *Conservation Biology* 20 (2006): 1–2.
13. M. Scott et al., "The Issue of Scale in Selecting and Designing Biological Reserves," in *Continental Conservation: Scientific Foundations of Regional Reserve Networks*, ed. M. Soulé and J. Terborgh (Washington, DC: Island Press, 1999).
14. K. Redford and J. Brosius, "Diversity and Homogenization in the Endgame," *Global Environmental Change: Human and Policy Dimensions* 16 (2006): 317–319.
15. D. Quammen, "An Endangered Idea," *National Geographic* (October 2006): 62–67.

16. R. Noss, "The Wildlands Project: Land Conservation Strategy," special issue, *Wild Earth* (1992): 10–25.
17. W. Newmark, "A Land-Bridge Island Perspective on Mammalian Extinctions in Western North American Parks," *Nature* 325 (1987): 430–432.
18. J. Scott et al., "Nature Reserves: Do They Capture the Full Range of America's Biological Diversity?" *Ecological Applications* 11 (2001): 999–1007.
19. G. Walther et al., "Ecological Responses to Climate Change," *Nature* 416 (2002): 389–395; T. Root et al., "Fingerprints of Global Warming on Wild Animals and Plants," *Nature* 421 (2003): 57–60; C. Thomas et al., "Extinction Risk from Climate Change," *Nature* 427 (2004): 145–148; J. McLaughlin et al., "Climate Change Hastens Population Extinctions," *PNAS* 99 (2002): 6070–6074; N. Dudley, *No Place to Hide: Effects of Climate Change on Protected Areas* (Gland, Switzerland: WWF Climate Change Programme, 2003).
20. T. Lovejoy, "Protected Areas: A Prism for a Changing World," *Trends in Ecology and Evolution* 21 (2006): 329–333.
21. C. Holling, "The Resilience of Terrestrial Ecosystems: Local Surprise and Global Change," in *Sustainable Development of the Biosphere*, ed. C. William and E. Munn (New York: Cambridge University Press for the International Institute for Applied Systems Analysis, 1986); C. Holling, "Cross-Scale Morphology, Geometry, and Dynamics of Ecosystems," *Ecological Monographs* 62 (1992): 447–502.
22. K. Brandon et al., eds., *Parks in Peril: People, Politics, and Protected Areas* (Washington, DC: Island Press, 1998).
23. R. Knight and T. Clark, "Boundaries between Public and Private Lands: Defining Obstacles, Finding Solutions," in *Stewardship across Boundaries*, ed. R. Knight and P. Landres (Washington, DC: Island Press, 1998).
24. R. Keiter, *Keeping Faith with Nature: Ecosystems, Democracy, and America's Public Lands* (New Haven, CT: Yale University Press, 2003); C. Sunnstein, "Beyond the Republican Revival," *Yale Law Journal* 97 (1988): 1539–1590.
25. Noss, "The Wildlands Project."

CASE STUDY ONE

# INNOVATORS DOWN UNDER: NEW ZEALAND'S FISHERIES

*Christopher M. Dewees*

As we stepped off the airplane on our first trip to New Zealand in 1986, my wife exclaimed, "So this is what clean air looks like!" This gorgeous island nation has a landmass about the size of California and a population of four million. We marveled at the unique plant and bird life that had evolved in isolation for millions of years before the first humans arrived from Polynesia less than a thousand years ago.

Why had I traveled six thousand miles to New Zealand, and why did I return again in 1995 and 2005? I wanted to study New Zealand's revolutionary experiment in marine fisheries management. New Zealand was the first place to privatize fish harvesting rights in an attempt to end the increasingly intense competitive "race for fish" that plagues most of the world's commercial fisheries.

Unlike many terrestrial natural resource endeavors (forestry, farming, ranching, mining) on private or government-managed lands, commercial fishing has long allowed anyone access to seafood resources. Imagine Iowa corn farming if access to land were not privatized. Farmers would develop the biggest, fastest tractors to outcompete others for the best farmland. You might even see dragster tractors! Soon farmers would dissipate all their profits competing to be the first to plant and harvest the most corn. This has long been the situation in commercial fisheries, whose resource is the last major wild food source harvested by "hunters."

A bit of history may help clarify the need for changes in the management of fisheries resources. Until the middle of the twentieth century, marine resources were treated as limitless. Government policies often encouraged development of fisheries with subsidies. As the human population grew and post–World War II technologies were adapted for increasingly efficient commercial fishing, nations realized that fisheries resources were finite and needed to be managed for long-term sustainability. Governments imposed regulations that limited harvest by restricting fishing gear, setting seasons, size limits, and annual catch limits; and finally limiting the number of vessels allowed to fish. Even with all that, the race for fish intensified dramatically. An extreme example was the Pacific halibut fishery, in which the approximately 50-million-pound quota was often landed or exceeded within a chaotic four-day season. The world was desperately searching for new ways to manage fisheries for biological, economic, and social sustainability.

In 1984, New Zealand's economy was in a serious crisis, and there was a strong call for immediate government action. The government quickly moved toward market-based systems, privatizing government services and assets, removing subsidies, and privatizing public lands. Marine fisheries, which were suffering from overfishing and poor economic performance, were swept along in the market-based policy revolution.

The combination of a national economic crisis, the collapse of many fisheries, and a new governmental market-based policy ideology created a perfect storm for change. The stage was set in New Zealand for rapid development and implementation of a comprehensive privatization of harvest rights with individual transferable quotas (ITQ). As one of the first nations to adopt ITQs as its primary fisheries management tool, New Zealand provides a fascinating case study.

After two years of design, the initial quota management system (QMS), using ITQs, was implemented in 1986 for all major finfish fisheries. New Zealand created ten fishery management areas (FMAs) and established total allowable catches (TACs) for each species within the FMAs. Permit holders were issued ITQs based on their 1982–84 catch history. Once established, the ITQs could be held in perpetuity, as well as bought, sold, or leased. When the government wanted to adjust the TACs up or down, it could buy back quota from quota owners or auction off new ITQs.

How did fishing industry participants react to this radically different management approach? During the first year approximately 20 percent of the quota owners sold out completely. In addition, as predicted by resource economists, the race for fish ended now that fishers were assured of access to their part of the total allowable catch and no longer had to compete with one another to catch fish.

The other immediate change was an increase in strategies by quota owners to add value to their catch through processing, market development, and chilled- or live-fish exports. Industry participants were obviously seeking to maximize profits from their limited quota holdings.

Additionally, New Zealand's First Nation tribes (Maori) had not been consulted or included in the initial allocation of harvest rights. The tribes initiated challenges to the QMS system based on the Treaty of Waitangi.

When we returned in 1995, New Zealand had made significant changes to the QMS. As the first country to implement a comprehensive harvest rights system, New Zealand had discovered some significant problems when putting this economic fishery management system into practice.

A primary discovery was the inefficiency of having government enter the quota market to adjust total allowable catches. The conservation incentives for industry were reduced because quota owners were guaranteed compensation for reduced quotas even if the reduction was due to their own overfishing.

To address that issue, New Zealand changed over to proportional ITQs. Each quota holder received a proportion of the overall TAC. That removed the government from the market and switched the economic and resource abundance risk onto the industry. Other nations learned from New Zealand's early experience, and the individual quota systems of those nations are proportional in nature.

A second major issue surfaced when the Treaty of Waitangi tribunal and the courts upheld the claims made by the Maori. Initial settlements of Maori fishery claims had occurred, and the Treaty of Waitangi Fisheries Commission was established to manage quotas on behalf of Maori. In 1992 the commission began a twelve-year procedure to determine the most equitable processes for allocating fisheries quota to Maori tribes and corporations.

Like the Maori, environmental and recreational fishing organizations had little involvement in the initial 1986 implementation of the QMS. That changed quickly. Environmental organizations scrutinized the government's setting of total allowable catches and sued the minister of fisheries when they disagreed with his TAC decisions. The environmental groups pushed the government toward a more precautionary approach when setting TACs.

Sportfishing groups increased their participation in the fishery management process by expressing concerns about allocation of fisheries resources between commercial and recreational fishing. This issue was especially intense for the snapper fishery in the densely populated Auckland region. The commercial industry perceived that the minister of fisheries was lowering their quotas to increase the fish available to the growing recreation sector.

When I returned to New Zealand in 2005, the quota management system was still evolving, as New Zealanders continued to address longstanding issues.

The nearly twenty-year process of addressing the fishery rights of the Maori reached a settlement with the passage of the Maori Fisheries Act in 2004. That began the allocation of quotas to individual tribes (*iwi*) or tribal organizations in perpetuity. It also established a corporation to manage several large seafood companies on behalf of Maori, with the dividends payable to the tribes. Maori now control over 50 percent of New Zealand's fishery assets.

Fisheries shared between user groups continued to be highly contentious. To partially address that, the minister of fisheries now sets a total allowable commercial catch after first subtracting recreational, Maori customary take, and other uses from the total allowable catch.

Environmental organizations are still involved in New Zealand fisheries. They have focused primarily on reducing the take of nontarget species, primarily seabirds and marine mammals. Their other focus is on the establishment of marine protected areas.

New Zealand exports about 85–90 percent of its seafood landings. In recent years the economic climate for fish exports has been poor due to the sustained strength of the New Zealand currency and the weakness of markets for high-end products. With these difficult economic conditions and the advancing age of fishers, many of the initial quota owners are retiring, and new entrants currently face a difficult economic climate.

There is much to be learned from New Zealand's pioneering approach to fisheries management. First, the economic theory of dedicated-access privileges appears to work in practice. The race for fish ended, and quota owners tried to maximize and sustain the value of their individual quota holdings. The primary goals of New Zealand's quota management system were to promote economic efficiency and industry profitability, "New Zealandize" the deepwater fisheries, maximize export earnings, and improve conservation of fisheries resources. On balance, New Zealand appears to have made good progress toward those goals during the first twenty years of its management system. This is especially true when one imagines how New Zealand's commercial fisheries might have performed if the QMS had not been implemented.

The benefits, however, come with costs and trade-offs. New Zealand's goals focused on economic and conservation improvements. Social goals were not stated, but three social issues quickly surfaced. Foremost were the treaty rights claims of the Maori tribes that had not been addressed in the original legislation. It took seventeen years from the initial Waitangi tribunal claims to get to the Maori Fisheries Act of 2004, which finalized settlement of those claims. This is a compelling example of moving First Nations' treaty claims into a modern economic context.

The second social issue is that of shared fisheries. New Zealand didn't specifically address the allocation of fisheries between the commercial, recreational, and Maori traditional uses in 1986. Though the government now sets a total allowable commercial catch, this is currently the most contentious and difficult issue within the QMS. Commercial quota owners feel that their quota assets (property rights) are at risk as other fishery sectors expand.

The third issue is that of aggregation of quotas by seafood companies. New Zealand set relatively high limits on consolidation (10–35 percent) to encourage the deepwater fishery development and maintain the pre-QMS history in which the largest ten vertically integrated companies landed two-thirds of the fish. Over time those companies have continued to consolidate quotas to the point that five holders now control 85–90 percent of all quotas. The obvious economic and social trade-offs in this situation are a topic of heated discussion in New Zealand.

As others around the world watch New Zealand's innovative approach, they should learn about what might or might not work in their own situations en route to designing a new generation of dedicated-access privilege systems to further improve fisheries management.

CASE STUDY TWO

# WHEN GOVERNMENT RESPECTS LANDHOLDERS: WILDLIFE IN ZIMBABWE

*Mike Jones*

Over a thirty-five-year period from 1975 until the government initiated its fast-track land reform program in 2000, Zimbabwean farmers achieved considerable success as custodians and beneficiaries of the wildlife on their lands. That success occurred because the government allowed farmers to use the wildlife that was on their lands in a manner that would ensure success for the farmers and conservation of the wildlife. The early successes achieved by farmers and the change of government from minority rule to majority rule in 1980 led to policy and legislative change that enabled communal land farmers to engage in wildlife-based enterprises through their district councils. The outcome of this policy reform was a fourfold increase in the amount of land available to wildlife, of which approximately one-quarter was in state-protected areas and the remainder was under either communal or private tenure.

In one sense, this is a story of macroeconomics and how change in a policy environment can create outcomes that are beneficial for wildlife and people. In another sense, it is a story of the ecology of wildlife and wildlife habitats in Zimbabwe, and a story of the enterprise, initiative, and growth of those who chose to look after the wildlife and make it the cornerstone of their livelihoods. Ultimately, what happened in Zimbabwe tells us what can happen when a government respects its people by treating them like responsible citizens. This story also reflects the crisis that occurred when the government chose to disrespect the people's right to land, resources, and livelihood and tells how some people found opportunity in chaos and forged new alliances and institutions to protect livelihoods and the wildlife on which they depend. The future of Zimbabwe's wildlife and people is still unfolding, and no one knows where it will end, but the prospects for a wildlife-based land economy are better than ever in some parts of Zimbabwe.

By the late 1950s, after a fifty-year period of increasing central control, Zimbabwe's wildlife was in decline. Attempts to increase wildlife numbers through the establishment of protected areas were negated by the widespread eradication of wildlife outside protected areas to control the spread of the tsetse fly and foot-and-mouth disease. Human populations were growing rapidly, wildlife habitat was converted to agricultural land, and ranchers were removing wildlife in large numbers because it competed for grazing and water. Many species were considered health risks to domestic stock, and large carnivores were regarded as vermin and destroyed. "One cannot farm in a zoo" was a commonly expressed sentiment. In 1960, the High Court ruled that natural vegetation was "property" and should be protected by shooting wild animals that were likely to eat it, dashing hopes that a protectionist policy would succeed in maintaining wildlife outside protected areas.

Government engaged in cautious institutional reform between 1960 and 1975 by encouraging landowners to try game ranching. Expansion of game ranching slowed toward the end of this period as subsidies for beef production increased and some ranchers shifted to safari hunting. Wildlife legislation promulgated in 1975 gave landholders a usufruct right to manage wildlife while it was on their land. That right gave landholders the authority, responsibility, and incentive to conserve wildlife, and it set the stage for the creation of a wildlife-based tourism industry and reversed the trends in habitat destruction and wildlife decline.

Wildlife legislation was changed again in 1982 to enable communal-land farmers to access some of the benefits from wildlife that their commercial farming neighbors enjoyed. Unlike commercial farmers, communal-land farmers do not have title to their land, which is under traditional (informal) rules of land tenure and subject to overriding control by local and central governments. The development of institutions and organizations for managing communal property required a different approach from that used to give rights to commercial farmers, who already had the right of land ownership as a basis for the grant of a wildlife usufruct right. Following some earlier experimentation with the donation of meat and payment of cash from the proceeds of elephant reduction operations inside protected areas, the government wildlife authority developed the Communal Areas Management Program for Indigenous Resources (CAMPFIRE) model in 1986. This allowed communities of farmers on communal land to have the authority to manage and benefit from wildlife.

In 1988 the Ministry of Local Government allowed two districts to manage wildlife on a trial basis using the CAMPFIRE legislation. The delay was due to disputes within the government about who should receive wildlife revenue—the producers, councils, or central government. The outcome was that government granted some wildlife-use rights to rural district councils, subject to conditions regarding wildlife revenue sharing between councils and the communities that produced wildlife. Unlike its view of commercial farmers, who were given a clear use right and considerable authority, the government's view of communal land farmers—and their ability to manage wildlife—included a strong element of self-serving paternalism. Farmers were expected to conserve wildlife in exchange for a small proportion of the revenue, while the authority and power were retained by the local and central governments.

Despite the shortcomings of the CAMPFIRE program, the governmental wildlife authority strongly encouraged it, so it came to include many areas adjacent to state-protected areas, where wildlife densities were comparatively high in relation to human densities.

From one experimental game ranch in 1959, by 1990 wildlife production emerged as a major component of land use, with 6.7 million acres (6.9 percent of Zimbabwe's land area) under commercial wildlife production. Another 8.4 million acres were added to that by 2000 through the expansion of the CAMPFIRE program. Overall, about 31 percent of the country, including state-protected areas, was used to produce wildlife and conserve habitat, which was a major shift from the trends that were apparent under command-and-control protectionism.

The areas set aside for wildlife production were mostly in the semiarid regions of the country, where variable rainfall of about twenty-four inches or less makes dry-land agriculture a high-risk occupation. Some farmers continued to grow crops and keep cattle, as well as produce wildlife. In the drier commercial farming areas, where crop production required irrigation and cattle production was marginal, some farmers converted entirely to wildlife. The switch to wildlife production increased due to the crippling effects of repeated droughts and decreased government subsidies for commercial farmers. In the higher-rainfall areas of the country, farmers integrated wildlife-based tourism and intensive breeding of species like sable and buffalo with agriculture, and it became an increasingly important additional source of on-farm income.

The outcome on communal lands varied considerably, depending on the degree to which councils devolved their authority to farming communities. Communal-land farmers generally practice subsistence agriculture and confine wildlife production to areas that are not heavily settled or cultivated. Farmers created specific wildlife areas in some districts, while councils reserved areas for wildlife in other districts, regardless of farmers' wishes. Crop raiding, particularly by elephants, was a major problem for communal-land farmers, who, unlike their commercial counterparts, lacked the firearms and authority to deal effectively with elephants. Various donor-funded fencing schemes were implemented but were largely unsuccessful because maintenance costs were high and fences were not a local solution to a local problem.

International prohibitions on the trade in species such as rhino and elephant were a constraint that further prevented farmers from receiving potentially significant income that could have been used to maintain wildlife habitats and provide the protection that these valuable animals need.

In 2000 Zimbabwe abandoned the rule of law under its fast-track land reform program, and the right to own land was taken from commercial farmers. Most commercial farmers were removed from their land, and the wildlife they once protected has largely gone. This fundamental breakdown of institutions necessary for economic development, which destroyed thousands of livelihoods, also created an opportunity for some farmers in the southeast of Zimbabwe.

Where commercial farmers have joined together to form wildlife conservancies, they are still on the land and wildlife production is their primary livelihood. The conservancies are building partnerships among themselves, with councils, and with communal-land farmers so that wildlife can be managed at appropriate scales and to increase the authority of communal-land farmers over wildlife and increase their wildlife income. Building these new institutions is contrary to current practice in the CAMPFIRE program, but the political will to change is occurring at the local level. Creating new institutions and new organizations will take time, but the prospects are high for wildlife to remain a primary component of land use in some parts of Zimbabwe.

The High Court decision that placed human rights to property and livelihood ahead of animal welfare was the single most important event in turning a failing command-and-control, preservationist approach to wildlife conservation into a highly successful national industry based on the principles of sustainable use. Zimbabwe followed that court decision with new legislation that empowered landholders and enabled them to become true stewards of the land and the wildlife communities that are part of that land. This is empirical evidence of Leopold's land ethic at work and testament to what farmers can do when government divests some of its authority to local landowners.

By comparison with their commercial farming neighbors, communal-land farmers were subject to the vagaries of government that put self-interest in wildlife revenues above the interests of the people it claims to serve. Despite that handicap, and despite constantly living on the edge of destitution, communal-land farmers have demonstrated a willingness to share their land with wildlife and suffer losses to crops and property that their commercial counterparts would not tolerate. There is a strong link between communal farmers, their land, and their wildlife that can be enhanced for the benefit of all, provided government is willing to treat communal-land farmers with the same respect as commercial farmers.

This story, which compares outcomes under two inequitable policy environments within the same country, demonstrates that command-and-control blueprints do not always protect nature and can be anathema to conservation, human rights, and economic development. In the complex social and ecological systems formed by people and land, a conservation system that enables flexibility and adaptability to local communities by devolving rights to landholders is the key to success. Our knowledge of adaptive cycles and the story of wildlife in Zimbabwe tell us that blueprints designed by central governments may fail.

CASE STUDY THREE

# KINZUA DEER COOPERATIVE: CONSERVATION THROUGH COOPERATION

*Jeffrey Kochel*

In northwest Pennsylvania one evening not too long ago, land managers, both private and public, gathered at a banquet with local businesses, state game commission personnel, outdoor writers, educators, and biologists to thank local hunters for their contributions to conservation by managing the whitetail deer population. Whereas it is not unusual for those people to meet on the issue of deer management, it is rare for them to gather in an atmosphere of respect and cooperation rather than polarized confrontation. The banquet was held in the Allegheny Plateau region of Pennsylvania, a region notorious for historically high deer populations and overbrowsed habitat. To appreciate why this example of cooperation is unusual, one needs to know the history of deer management in Pennsylvania, its ecological effects, and the historic relationships between hunters and landowners.

The Pennsylvania Game Commission, supported primarily by revenue from hunting license sales, is the state agency responsible for deer management. It was established in 1895 with the primary responsibility of reviving a nearly nonexistent whitetail deer herd. By 1928 the commission had been so successful that the herd had rebounded and begun overbrowsing its habitat. As a result, the first statewide doe harvest was conducted. It met with significant opposition from politically active hunters and local newspapers and began nearly eighty years of conflict between hunters and their allies and land managers such as foresters and farmers. Hunters were supported by rural merchants, tourism associations, and rural legislative representatives, while foresters and farmers had few supporters. Deer populations fluctuated somewhat during that period, with a few brief declines. Each time the population declined enough to have a positive impact on habitat, however, political pressure by hunters and their supporters caused the game commission to reduce antlerless harvest permit allocations. This allowed the herd to grow to levels beyond the habitat's carrying capacity. Years of conflict over this issue resulted in two opposing views:

- Hunters were interested in high deer numbers and believed claims of habitat degradation were exaggerated and unrelated to deer overabundance.
- Land managers and farmers realized that overbrowsing by deer was affecting their lands, but they were politically marginal and lacked data to support their claim. They assumed hunters were not interested in conserving habitat.

Hunters and land managers refused to communicate, and once their opposing views became entrenched, it was natural that anyone with a stake in the issue would choose sides rather than look for solutions.

The Allegheny Plateau is characterized by a mix of private and public timberlands providing a relatively unbroken forest cover across the plateau. Within these ownerships is a wide range of stand characteristics, all with one thing in common: They have been severely affected by overbrowsing. Large areas of mature timber have understories dominated by ferns, grasses, beech, and striped maple. These species all have low browse value, yet the beech and striped maple have been so heavily browsed they rarely get above waist high. Understory diversity is extremely limited, and there is little seedling regeneration throughout the region.

In an attempt to deal with overbrowsing issues, some managers began using herbicides to eliminate the dense understory of undesirable vegetation not preferred by deer. Woven wire fencing was also used to exclude deer from regenerating forest areas. These practices addressed regeneration of desired tree species only in localized areas and at great expense, while most forestlands were being altered or lost. Landowners also attempted to work with hunters to improve doe harvests. By opening their lands to public hunting and providing maps and access, they hoped to encourage a greater harvest. Others used hunting leases in an effort to promote doe harvesting. Despite those efforts, results were localized, erratic, and dependent on antlerless harvest allocations by the game commission, an agency with limited resources and captive to its funding constituency, the hunting license buyer.

What changed that allowed the opposing sides to come together and work cooperatively on solutions to these issues? First, scientific evidence became available that indicated the dramatic negative effects of overbrowsing by deer. The research was led by the USDA's Northeast Forest Research Station (NFRS). Second, conservation organizations and community groups joined with farmers, foresters, and other landowners in calling for a revision of the game commission's deer management strategy to a more habitat-based approach. Third, a group of forest managers and scientists who had been working together for decades on deer management issues recognized that there might be a window of opportunity to make a real difference in reducing deer numbers, improving habitat conditions, reducing regeneration costs, and increasing forest productivity. Finally, the Sand County Foundation (SCF) served as a catalyst to help address the issue of overabundant deer and depleted habitat. SCF had developed a program called Quality Hunting Ecology, which espouses the philosophy that proper management of deer and associated habitats is essential to ensure the health of both.

In 2000 this group of land managers, the NFRS, and SCF joined to develop a blueprint for bringing people together, including hunters and local businesses, in an effort to improve deer and habitat management. The landowners included two sawmills, a timber-investment management organization, a municipal watershed authority, and a district of the Allegheny National Forest.

The desire of the land managers combined with the approach of SCF was a perfect match that resulted in the formation of the Kinzua Quality Deer Cooperative (KQDC). Although it encompasses five different ownerships, it is managed as a single 74,000-acre deer management unit. Its objective is to apply the principles of SCF's Quality Hunting Ecology over a large area to develop a long-term demonstration project. The project goals are to create, improve, maintain, and monitor habitat that is more diverse, productive, and capable of sustaining a healthier deer herd while maintaining hunter satisfaction. This is accomplished by working in partnership with a broad range of stakeholders, with a special emphasis on sportsmen as an integral part of the conservation effort.

The five land managers constitute the leadership team for the KQDC and are responsible for providing support, approving activities, and, along with SCF, guiding the overall project. The leadership team receives input from and works closely with all the other participating cooperators. All the participants have a stake in improving the habitat, increasing deer quality, and maintaining hunter satisfaction. The project is generating data and educational information that are used by the leadership team in its management planning activities. This has become a truly bottom-up approach driven by local interests working in cooperation with state and federal agencies.

KQDC deer monitoring activities include annual surveys of deer and land health. Data on deer density and abundance, fawn recruitment, and herd age and sex ratios are collected. This information is combined with hunter surveys. The data are summarized and published on the KQDC Web site (www.kqdc.com), allowing open access to the information.

The success of this project is dependent on engaging hunters so that they realize they are a key part of the conservation process. The banquet mentioned in the beginning is an example of that. Outside speakers are brought in to help reinforce the importance of healthy habitat and help shift the thought process from number of deer to health of the deer habitat.

Today, positive changes are evident throughout the KQDC area. Commercial-size tree seedlings are evident where none were visible five years ago (figure C3.1). The traditional timber harvest areas that appeared to be lost are regenerating, and there are signs throughout the area of a recovering habitat.

Between 2000 and 2006 sample plots showing no deer browsing increased by 100 percent, and plots that lacked seedling regeneration decreased by 10 percent. The deer population index has declined by nearly 50 percent. The ratios of fawns to does and bucks to does have increased. Bucks in every age class have seen average antler spread and number of antler points increase. Average body weight for deer of all ages has increased. Forest regenerating in harvest areas without fences has increased from 270 acres to over 1,300 acres. These number all suggest a recovering forest habitat and a healthier deer herd.

A key element that helped facilitate these improvements was the institution of more liberal and innovative deer harvest regulations by the Pennsylvania Game Commission. In 2001 the commission moved from a three-day antlerless season to a two-week concurrent antlered and antlerless season. In 2002 three-point antler restrictions were implemented, and in 2003 a Deer Management Assistance Program was developed to enable landowners with demonstrated deer overabundance to obtain additional doe harvest allocations. These changes all worked to benefit the KQDC.

Figure C3.1 A healthier forest on the Kinzua Quality Deer Cooperative, Pennsylvania.

Regulations aside, the fundamental shift has been from top-down state regulations to more local decision making.

Hunters are still skeptical and complain about the perceived lack of deer, but an increasing number are beginning to comment that seeing fewer deer is worth the opportunity to harvest a large buck or even a large doe. In fact, some hunters have joked that we should no longer promote the KQDC because they want to keep the quality of the deer in that area a secret!

Despite the success of this project, several threats exist. The game commission is under continual political pressure to increase deer numbers; that could cause them to reduce antlerless permit allocations at any time and would result in the herd returning to unhealthy levels. The number of hunters in Pennsylvania continues to decline, while the mean age of hunters continues to rise. In addition, there are still many hunters who threaten to stop hunting unless deer numbers increase; that could result in too few hunters to control deer populations. Finally, and ironically, the more successful this project is, the more difficult hunting becomes. As habitat improves, regenerated areas develop faster and older stands have greater structural diversity, making hunting more difficult and reducing hunter success rates and satisfaction.

These new problems are a welcome challenge, however, and the Kinzua Quality Deer Cooperative is the ideal place to address them.

CASE STUDY FOUR

# WORKING WILDLANDS

*Lynne Sherrod*

Small clouds of dust sifted across our boots as we walked the winding road through Sam Capps's mountain heifer pasture. Sam, a quiet and intensely personal man in his late eighties, was shyly but in great detail explaining the complexities of his impressive breeding program. He was warming to his topic, much as the midmorning sun was warming to the day.

A profound sense of peace flooded through me as I surveyed the hushed splendor that stretched out for miles before us. Since childhood I had tried to envision what the land looked like when the first settlers arrived. I was raised on stories of how my family had homesteaded in northwestern Colorado five generations earlier. Every Memorial Day my cousins and I were marched through the Yampa Cemetery to better appreciate our family legacy. Our parents proudly pointed to the row of stones that paraded back to a post–Civil War era when settlers flooded the West to start new lives in exchange for hard work and determination.

Absorbing the early fall glory that was staining the countryside with a patch quilt of colors, I suddenly stopped in my tracks as Sam pointed out a tall stand of ponderosa pine and said it had been there before his grandfather homesteaded this ranch following the Civil War. He observed that someone in his family had watched those trees grow and thrive for well over one hundred years.

For an instant I wondered if I had unwittingly shared my mental meanderings out loud or if Sam were a mind reader. Again, I was reminded of the common bond that people of the land share—a deep sense of belonging, with a pervading gratitude and respect for those previous generations whose hand-hewn cabins dotted the West. Determined wills and strong backs had thus far protected these bountiful landscapes, along with an immeasurably rich human and natural heritage. They had left us something tangible with which to continue shaping the future for the generations yet unborn. Still emanating from the past was a value centered in a cultural ethic that inspired hope, exacted discipline, and fostered belief in something greater than one's self.

Listening to Sam as he shared the precious gems of his birthright, I was struck by how the past is interwoven with our future. Our time on earth is but a vapor, and it is understood that we are responsible for stewarding these resources. What we leave behind is a gift for the next generation. Will the cycle continue to repeat itself, or will ours be the last generation that passes on this legacy?

Sam's answer was obvious as he pointed out the pasture where he cultivated special grasses designed to supplement the flourishing elk and deer populations that also called his ranch home. We paused to study a dying musk thistle along the road, and he excitedly pointed out the successful deployment of a biological army of bugs that he had ordered to help him wage war against this invasive interloper.

Sam had a vision for his ranch, one that had been instilled in him with every step he took as he followed in his father's and grandfather's boot prints (figure C4.1). During the Great Depression much of the original ranch had been lost, so he and his father committed themselves to restoring their original operation. With single mindedness and spartan determination, he had eventually surpassed that goal and was now showing me a part of the resulting 28,000-acre ranch. Several lifetimes of accrued

Figure C4.1 Sam Capps moving cows on his ranch in Colorado.

knowledge and stewardship practices were evident in the caressing glance with which Sam regarded the view before him.

Truthfully, this land was the heart and soul of his existence—by his measure, the sum total of his purpose on this earth. A bachelor, with no relatives he cared to leave the ranch to, he was facing his own mortality and was adamant that the ranch should continue to thrive long after he was gone. He was unwavering in his insistence that it not fall victim to the tentacles of sprawl that he saw inching their way south along the Front Range of the Southern Rockies, drawing ever nearer to his lifetime home and the ranch he loved.

Sam's determination sent him on a quest through the myriad complexities of estate planning, which ultimately led him to a revolutionary device—a legal document that could guarantee the permanent protection of his cherished land, forever holding unwanted development at bay. But the words "conservation easement" conjured up images of environmental constraints governed by interests that had little appreciation or understanding of production agriculture.

Sam finally found the answer he was looking for in the Colorado Cattlemen's Agricultural Land Trust (CCALT), the first land trust in the country to be founded by a group of mainstream livestock producers. CCALT was formed by ranchers in the mid-1990s to provide like-minded individuals with the comfort of embarking on this perpetual partnership with peers who understood that, along with a bounty of conservation values, a lifestyle and business must also be safeguarded.

Sacrilege or genius? One astounded environmentalist observed that cowboys and conservation were oxymoronic, while some in the livestock industry thought those ranching visionaries had gone over to the dark side. Still others would say that participating landowners were attempting to rule from the grave. Were they merely forestalling the inevitable? Sam didn't think so.

For me, as a representative of CCALT, there to talk with him about his dreams for the future of his ranch, meeting Sam was one of the privileges of a lifetime. He embodied the image of thousands of ranchers who lived out the edicts of their hearts. Such stewards of the land had cultivated and were continuing to nurture the natural resources that all society values and ultimately benefits from, not to mention the all-important production of food and fiber.

According to the U.S. Fish and Wildlife Service, over 70 percent of the wildlife that thrives in the vast open spaces of the West lives on private lands. Those working wildlands safeguard plant and animal communities while keeping intact the broad expanse of scenic views that the West is known for and that regularly draws tourists from far-reaching locales. Private ranchlands host a variety of recreational opportunities but also provide an increasingly important buffer to encroaching sprawl and its detrimental impact on the natural environment.

Science tells us that the half of the West that is private is the most biologically productive: Those lands occur at the lowest elevation, have the deepest soils, and are the best watered. Indeed, many of the original homesteads and ranches of today lie along the fertile bottomlands of valleys and corridors, serving as society's last stand against the pervasive growth that is now engulfing the boundaries of our public lands. If these remaining ranching operations convert to housing developments, the last line of defense protecting our public land boundaries will have fallen.

In the past these private working wildlands were taken for granted, viewed as limitless. With the rapidly increasing fragmentation of our wide-open spaces—our watersheds, wildlife corridors, natural areas, viewscapes, and agricultural lands—it seems as though the West now has one big "For Sale" sign protruding from its very heart. It has become obvious that we either embrace new solutions to this assault on the life-beat of the land or stand by rooted in powerless sorrow and watch it slowly ebb away until nothing even remotely resembling a healthy ecosystem is left to mourn over.

Growth isn't bad when carefully managed, and certainly there are rural communities across this country that would give their eyeteeth for just a little development pressure. Since Horace Greeley's admonition to "Go west, young man, go west," this big-hearted place has welcomed all comers. Do we now just slam the door in their collective faces? How do we encourage sensible development in areas without permanently risking the wonders that initiated this dilemma in the first place? Can the government afford to buy and manage all the land that needs protecting? What's the alternative?

Private land stewards like Sam have been capably managing these working landscapes for generations, bearing the bulk of responsibility on their own hardworking shoulders and limited bank accounts. CCALT believes that if you offer these folks a little say over their future, they might be interested in what was formerly perceived to be the favored tool of the "green" persuasion.

The very term *conservation easement* is often enough to give a landowner pause, especially if it's being bandied about by someone whose intentions are suspect. First, it implies that someone other than the landowner is defining conservation; and, second, and even more disquieting, it implies public access.

In addition to permanently conserving his land in a way that guaranteed that his stewardship could continue, Sam wanted to enable his chosen heirs to inherit his estate. He had formed a partnership with his ranch manager of thirty years, and the easement became an important tool in not only preserving his own heritage, but also ensuring that the manager and his family could maintain Sam's legacy and reducing the value of the estate so that they could afford that privilege.

As Sam and I continued down that dusty road, I asked him why he cared what happened to the ranch after he was gone. It was important to understand each landowner's motivation for considering such a binding choice. Forever is, after all, a very long time. His answer, framed against the backdrop of a deep-blue Colorado sky, will forever be imprinted on my soul and encapsulates the very definition of a land ethic. With fists unconsciously clenched, he hesitated and looked me straight in the eye for a long moment as if weighing my worth; then in a low and intense voice he firmly stated, "Because I want to know that this land always looks the same as the day God made it."

As he abruptly turned and resumed his brisk pace up the road, I felt as though he had just placed a priceless family heirloom in my hands for safekeeping. He reminded me of my husband, whose love for the land was knit into the very fiber of his being, although articulating that passion was not a skill he was interested in displaying. Chancing another question, I asked Sam what his neighbors were likely to think. "I expect they'll think I'm crazy," he shared with a wry smile, "but I hope they'll do the same thing."

Sam needn't have worried. His leadership, and the eventual passage of a statewide tax incentive, inspired many of his neighbors to follow suit. Sam is now gone, but I am sure he'd be pleased to know that his neighbors have protected an additional 35,000 acres, with a total of 100,000 acres in the surrounding vicinity. With the fate of the Capps Ranch secure, his neighbors understood that they had a chance to preserve the integrity of their valley and the place they all called home. Whoever said that one person couldn't make a difference?

Cattlemen's land trusts have proved to be an important mechanism for preserving agricultural opportunities and private land stewardship while protecting the open space and wildlife habitat so valued by western residents and visitors.

In 2004 CCALT joined other state cattlemen's associations that had followed suit by starting their own land trusts to form the Partnership of Rangeland Trusts (PORT), a collaboration between locally based, agriculturally oriented groups that work with private landowners on voluntary conservation programs in the West.

By January 2007 the seven founding members (and more members on the way) had partnered in the protection of over 1.2 million acres, representing 1 out of every 6 acres of private land permanently secured by conservation easements in the entire country.

The success of Sam's vision continues to be defined by CCALT's original guiding principle: Organizations can't protect land; only people protect land.

# PART II

## *A Changing Toolbox for Conservation*

Part II examines a number of new approaches to practicing natural resources conservation. In chapter 5 Jodi A. Hilty of the Wildlife Conservation Society and Craig R. Groves of The Nature Conservancy present a comprehensive picture of conservation planning. Planning may be a bad word in the minds of some, but in the absence of appropriate planning, and implementation of plans, conservation is likely to meet its targets as often as a blue moon appears (not often). In other words, natural resources management should be taken seriously, as money and human capital are always in short supply. In order to ensure that conservation actions are thought out, employ appropriate ideas and technology, and are designed so management actions are done adaptively, planning is essential.

Chapter 6 is by Lora A. Lucero, a land-use planner in Albuquerque, New Mexico. Some might wonder what role community planning plays in natural resources management. Given that 25 percent of private land in the contiguous United States is in exurban development and sprawl, the need for livable towns and cities is urgent. Conserving nature and ecosystem services requires workable planning to build sustainable urban communities for people; sprawl is not the answer ecologically, economically, or culturally.

Chapter 7 was written by conservationists associated with the Sonoran Institute, headquartered in Tucson, Arizona, with offices across the West.

The increasing rate of development of open lands and the parallel increases in the value of real estate make it prohibitive for anyone to purchase all the important private conservation lands. This chapter examines incentives for landowners and developers to integrate conservation into their projects and land uses so that development, in essence, contributes to conservation.

J. B. Ruhl, a lawyer at Florida State University, examines the value of ecosystem services in chapter 8. Flood control, soil stability, pollination, regulation of atmospheric gases, and the aesthetic and recreational values of open lands are examples of services that healthy landscapes provide. Historically, we have taken those services for granted, but an increasingly crowded planet and the degradation of lands and waters are causing people to appreciate how much it costs to pay for substituting those services once they are gone. For example, healthy wetlands can buffer and absorb the energy from hurricanes that threaten millions of people living in coastal cities. Engineering infrastructures to substitute for the role of those wetlands when they are replaced with development is extremely costly (and as Hurricane Katrina demonstrated, our engineered solutions may not always suffice).

Part II concludes with three case studies, each demonstrating novel economic approaches to conservation and natural resources management. William J. Ginn works for The Nature Conservancy in New England, Chris Kelly works for the Conservation Fund in northern California, and Kay and David James ranch near Durango, Colorado. Each of the contributions tells a story of conservation and natural resources management that work economically, as well as ecologically and socially, whether protecting forests or raising grass-finished cows on healthy rangelands. These case studies also offer examples of lands that are being kept open and out of development while simultaneously maintaining their productive capacity—in other words, of working wildlands.

CHAPTER FIVE

# CONSERVATION PLANNING: NEW TOOLS AND NEW APPROACHES

*Jodi A. Hilty and Craig R. Groves*

It was 2003 when the Wildlife Conservation Society was asked to lend its help to conservation efforts in the Madison Valley of southwestern Montana. A group of ranchers had formed the Madison Valley Ranchlands Group (MVRG) with the goal of protecting their ranching way of life and the open spaces on which ranching depends. They sought to engage citizens in making the decisions about the future of the valley that could affect livelihoods, wildlife, and ecosystem processes. Part of MVRG's organizational strategy was to develop partnerships with other organizations that had the capacity to provide information and help to achieve shared goals. The Wildlife Conservation Society became a partner with a common goal of designing and implementing a wildlife conservation assessment for the valley (http://www.wcs.org/international/northamerica/yellowstone/settingprioritiesforconservation).

Various groups in the valley that are associated with MVRG are now using information from that assessment to drive their collective decision making around balancing wildlife priorities and human development. For example, a citizen-based wildlife committee is addressing a variety of wildlife management issues, while another ad hoc group is focused on rural residential sprawl. In short, a cast of conservation-oriented stakeholders is collectively engaged in the planning process and is slowly taking steps toward implementation.

The collection of partners working with MVRG was a mixed group of local residents; environmental nonprofits; county planners; the state wildlife agency; and federal natural resource agencies, particularly the U.S. Forest Service and the Bureau of Land Management. The partnerships and processes in Madison Valley reflect recent shifts in natural resources management: (1) to planning across rather than solely within administrative boundaries, (2) to early and broad stakeholder engagement, and (3) to bottom-up planning compared to top-down approaches. These shifts, along with the emergence of new tools such as remote sensing and spatial analysis and disciplines such as landscape ecology, enabled the planning efforts to move forward, always with an eye toward implementation.

* * * * * *

With habitat loss and fragmentation continuing at an unprecedented pace and an earth already widely impacted by human activities,[1] we need to target conservation efforts toward those natural places that are most intact and biologically important. If habitat loss and related habitat fragmentation are the largest threats to conservation of biodiversity, a host of other threats compounds them. Depending on the location, those threats could encompass the invasion of exotic species; human alteration of fire, flooding, and other natural processes; human-induced climate change; and rural sprawl, the fastest-growing land use in the United States.[2] Yet social, political, and economic support for conservation is inconsistent and arguably insufficient to protect our remaining natural heritage. This situation makes it all the more important to develop plans and priorities with which to allocate conservation resources.

How do we prioritize what ecosystem processes, species, and landscapes merit our attention, and where we should make our conservation investments? The concept of conservation planning was developed in the early 1970s as a methodology for conserving biodiversity, an approach that could occur at many different spatial scales.[3] Many countries and groups subsequently began to focus on conserving an ecologically representative set of protected areas as a response to biodiversity loss. For example, Canada is working at national and provincial scales to identify and establish new protected areas in designated regions on the basis of ecosystem representation and integrity. As the field of conservation planning has developed, so too has the focus of efforts—from early plans that emphasized species conservation to a greater emphasis on ecosystems, processes, and even ecosystem services as conservation targets.[4]

While the purpose of conservation planning is to define where and how conservation actions should occur, the various approaches that have been developed do not necessarily lead to the same on-the-ground priorities. This discrepancy is largely due to differing agendas (e.g., priority ecoregions versus biodiversity hotspots) and data that are used. For example, a review of the strategies of thirteen conservation organizations revealed twenty-one different approaches to prioritizing targets for conserving biodiversity.[5] In general, conservation planning has emphasized where conservation should occur and focused less on how it can be achieved.[6] Indeed, some scientists suggest that the where and how of conservation are really two separate fields of conservation science.[7] Conservation assessments define where conservation should occur, and conservation planning uses assessments to develop an implementation strategy.

Today, a variety of entities engage in conservation planning. Federal agencies, such as the U.S. Forest Service, undertake planning at scales from regions to watersheds. State agencies use conservation planning principles, as well. A recent example is the comprehensive effort achieved by all states to create state wildlife action plans (http://wildlifeactionplans.org/), which comprise each state's efforts to prioritize conservation actions for wildlife species and habitats. Conservation organizations, from land trusts to community-based programs and science-based organizations, use conservation planning to help prioritize where to place their resources.[8] Land-use planning by local government often incorporates similar principles. Arizona's Sonoran Desert Conservation Plan in Pima County is exemplary in that regard (http://www.pima.gov/cmo/sdcp/intro.html). While no process is identical to any other, there are some basic components of conservation planning efforts that help ensure success. We will review those below, but will first offer a brief overview of the emergence of the discipline of landscape ecology and the tools used in conservation planning.

### Emergence of Landscape Ecology

Our ability to do conservation planning at the appropriate spatial scale was previously limited by a lack of tools. Today, with the development of new tools and advances in landscape ecology, we can examine multiple spatial scales. Landscape ecology is largely concerned with the interaction among different community types, compared to traditional ecological inquiries that focused on single communities or even populations within a community.[9] Landscape ecology is a discipline that hierarchically layers individuals, populations, communities, and ecosystems at different scales, each level of biological organization possessing properties not featured by the levels below it (see table 5.1 for examples of emergent properties).

One shift in scientific inquiry that is attributed to landscape ecology is the increased emphasis of planning across boundaries. Boundaries include not only biological community types, as mentioned above, but also geopolitical boundaries delineated by humans. Today increasing numbers of studies and conservation efforts span multiple jurisdictions and management regimes. This new approach has helped contribute to a paradigm shift in the conservation community from a predominantly public lands

| Table 5.1<br>Some Emergent Properties of Landscape Ecological Systems |
|---|
| Types of communities present |
| Diversity (numbers of patches and proportions) of community-types |
| Spatial extent (and proportions) of various community-types |
| Biomass (by community-type and overall) |
| Spatial configuration (dispersion, sizes, shapes, juxtapositions) of patches |
| Ecotonal features (edge effects) |
| Connectedness (links among patches, corridors, barriers) |
| Interpatch fluxes (energy, nutrients, organisms, information) |
| Stability (resilience, constancy, predictability) |
| Long-term trends (succession, degradation, exotic invasion, extinctions) |
| Energetic properties (productivities, decomposition rates, turnover times, trophic efficiencies) of individual community-types and the entire array of communities |
| Historical context |
| Anthropogenic influences |

focus to a focus on the public-private land interface and private land conservation (although land trust organizations have long focused on private land acquisition). Working across boundaries enables scientists and managers alike to address challenges like the cumulative impacts of human activities within a given landscape or management of species across their whole range. Such opportunities facilitate more consistent management across boundaries, although that is dependent on willing land managers and stakeholders.[10]

### Importance of Spatial and Temporal Scale

Both spatial and temporal scales affect the way we understand a landscape and conduct conservation planning. At the heart of landscape ecology is our ability to use data at many spatial scales. The scales we choose will affect how we think about conservation planning, including our perception of landscape patterns and processes. Ideally, the goals of the conservation plan will determine selection of the appropriate planning scale. The goals may be for one or more populations of a species, a species across its range, communities, or watersheds, or they may focus on ecosystem conservation. Failure to be explicit about which elements of biodiversity we are trying to conserve and at what scale we are planning may inadvertently result in inadequate efforts for some species or processes.[11]

Spatial scale makes intuitive sense, but terminology can be confusing, so we will briefly review a few important terms and concepts. In particular, we use *coarse* and *fine* scale to distinguish differences in the *grain*—the level of detail of the data—and *spatial resolution* (also known as extent) to describe how large or small the area of focus is. For example, conservation planning at a coarse scale and large spatial extent may encompass the entire distribution of a species, whereas a finer-scale effort may focus on a specific population or daily travel corridor. In general, the trade-off conservation planners face is that the larger the extent of a planning region, the coarser the grain of data will be. For example, a watershed plan may use detailed aerial photography to map land cover, whereas a statewide conservation plan would likely use coarser-grain satellite imagery for the same purpose (figure 5.1).

The resolution of the data available may limit how fine the scale of an analysis may be. It is critical to understand that our perception of landscape patterns is dependent on the resolution of data available.[12] The geographical extent—that is, the area chosen as a focus for planning efforts—will determine the type of data that can be used and also influences the outcomes of any plan. This is because the geographical extent determines what biotic and abiotic processes are included.[13] Whereas the challenge with digital data can be the limitations of using coarse-scale data on smaller regions, we often have the opposite challenge with ecological data collected in the field. Many field studies measure ecosystem variables in a small spatial extent. The difficulty and inherent danger is then scaling up that information to be applicable to a larger area that almost certainly has greater geographical and ecological variability.

Conservation planning may occur at a wide range of scales, although most planning efforts are at a scale of one to a few hundred square miles. Some conservation planning efforts have emerged that span entire countries, and there are even transboundary multicountry efforts (e.g., the Yellowstone to Yukon Conservation Initiative).[14] Whereas local- and regional-level planning may focus on specific biological features such as focal species, plans for broader regions are likely to focus on the intactness of communities and ecosystems rather than specific species. The reality is that implementing plans across large expanses still generally requires a series of localized efforts across the region.[15] Our interest in the Madison Valley originated because a coarse-scale analysis pointed to the valley as important for the region's biodiversity and for keeping the Greater Yellowstone Ecosystem linked to other core wilderness complexes, such as in central Idaho.

Temporal scale is also important to consider. Most ecological studies are conducted in less than five years, but ecological processes such as fires and floods that can transform landscapes occur over much longer time periods. The significance of this is that planning should be based not on the presence of a species at a specific point in time, but rather on a more robust approach that allows for ecosystems to change over time, and should still incorporate a strong likelihood that any focal species will maintain viable populations given the planning scenario. Failure to be explicit about the period of time over which a conservation plan should function and to allow for dynamic changes that occur over time may mean that the plan will not be successful.[16]

## A Review of Conservation Planning Components

We have presented an overview of how the field of conservation planning came into existence and some of the new approaches available to support

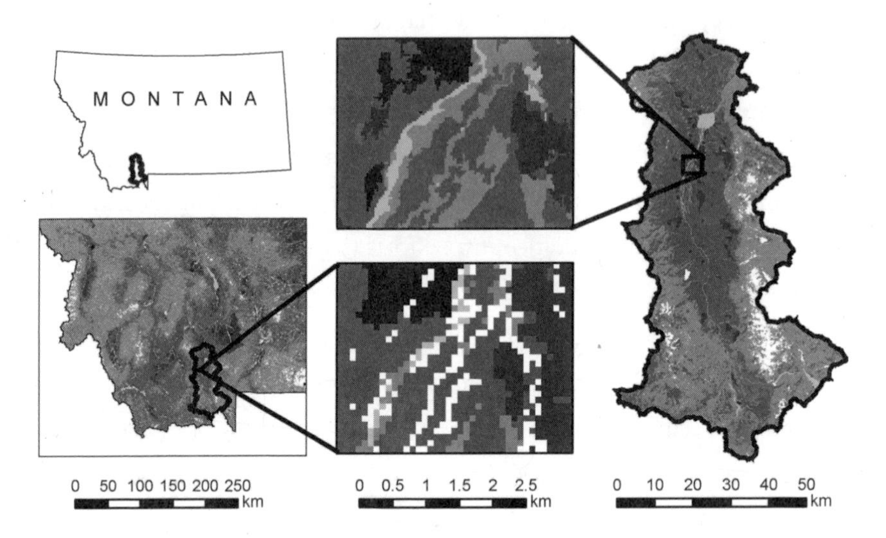


Figure 5.1 The Madison Valley in Montana is represented on the left by approximately 90-meter (~ 90-yard) GAP (Gap Analysis Program) data scaled at 1:10,000,000. The Madison Valley figure on the right is approximately 1-meter (~ 1-yard) SILC (Satellite Image Landcover Classification) land cover from the Wildlife Spatial Analysis Lab at the University of Montana–Missoula) scaled at 1:1,750,000. The center maps are blown up to the same 1:100,000 scale, illustrating the scalar differences.

the process. Next, we will offer a review of the typical steps that have proven to be successful in project-level conservation planning. These steps assume that the regional-scale question of where action should be taken has been addressed and that the focus for the project-level effort is on developing on-the-ground action strategies.[17] The following general steps of site-based conservation planning, based on the planning methodologies of The Nature Conservancy and the Wildlife Conservation Society (figure 5.2),[18] guided our efforts in the Madison Valley:

- *Defining the goals and objectives.* Failure to define goals initially will lead to a lack of clarity throughout the process and perhaps even project failure. Defining goals includes selecting targets (species, processes,

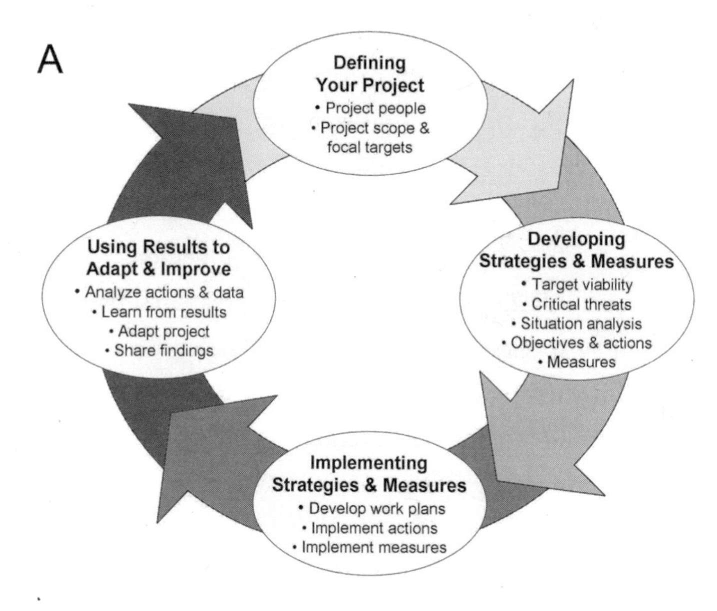

Figure 5.2 The Nature Conservancy's Conservation Action Planning approach (A) and the Wildlife Conservation Society's conservation planning framework (B) outline similar components needed to achieve site-level conservation plans.

communities or ecosystems), specifying the desired states of those targets, and identifying and prioritizing what threats the plan will address. It is critical to define the explicit targets that a successful plan would conserve. For example, focal species are often chosen to serve as surrogates for other elements of biodiversity and may be keystone, umbrella, flagship, indicator, specialist, or vulnerable species.[19] The idea is that by selecting and designing a conservation strategy for these species, other species will also be conserved because the focal species have the broadest requirements. While a focal-species approach can be an important tool for conservation, other biophysical targets may be important to conserve the entire ecosystem on which individual species depend.[20]

- *Planning and prioritizing actions.* Here the goal is to determine priority areas for the target species in relationship to where the threats occur in the project area. An often important component of these analyses is an assessment of the costs of abating threats at locations across the landscape. Understanding where the current and future potential threats are on the landscape and the amount of resources needed to address those threats helps in selecting priority areas and defining what actions should be taken.
- *Implementing and monitoring action.* An important and often skipped step after creating a strategy is developing a means of measuring success. How do we know whether our goals are being achieved? It is difficult on limited budgets to imagine implementing, measuring, and monitoring. There are a few good examples of ongoing efforts to address this challenge. The Conservation Measures Partnership represents a collection of nongovernmental organizations working together to develop standards to design, manage, and measure the success of conservation projects. One tool they have created is the Open Standards for the Practice of Conservation (www.conservationmeasures.org/CMP/).

- *Reporting, reviewing, and adapting.* The follow-up to monitoring is analyzing the collected data to make sure the actions are moving the project toward desired outcomes. If not, managing adaptively may require developing new actions or strategies to attain the project's stated goals.

An important component of conservation planning is connecting to conservation networks in the region surrounding the area of focus. With the ever-increasing amount of habitat loss, conservation areas in isolation are often inadequate for retaining biodiversity.[21] The concept of conservation area networks is that each conservation area or core area is connected to other core areas, through one or more movement corridors, for example.

Arguably the most essential element to incorporate into conservation planning is the involvement of constituencies. In reality, a plan without a supportive constituency for implementation is likely to be relegated to a dusty shelf. The involvement of local stakeholder groups such as the Madison Valley Ranchlands Group can be critical. At the very initiation of a project it is important to conduct a stakeholder assessment and identify the primary and secondary stakeholders as well as any obvious groups or individuals that may be opposed to the project. Inclusion of social scientists helps in the understanding of stakeholder attitudes, values, and actions that may affect the successful implementation of any plan. Ultimately, a successful plan will integrate both social and biological realities into any outcome.[22] In fact, one means to improving conservation models is to ensure that they are interdisciplinary.[23] Additionally, stakeholders are more likely to participate and assist in the process if they are kept apprised of the project's progress and have opportunities to provide input.[24] Delivering scientific and conservation information in useful formats to nonscientists and nonprofessional conservationists will be important so that all stakeholders can use the output information.[25]

## Challenges Faced by Conservation Planners

The challenges of conservation planning span a wide set of issues. Methodological issues exist, despite advances in the field, that limit our effectiveness. Also, socioeconomics drives both opportunity and the limitations of resources available for conservation planning.

### *Methodological Challenges*

Arguably the greatest challenge for conservation planners is incorporating the dynamic nature of ecosystems and ecological processes and not managing for a single state. Maintaining a dynamic ecosystem generally requires maintaining ecosystem processes. Whether ecosystem processes should be prioritized in conservation plans has been the focus of considerable debate. Some scientists argue that too much focus on process could lead to overlooking the elements of the system and that processes may continue unimpeded despite a biotic transformation such as the introduction of an exotic species.[26] Others argue that as processes become compromised, biodiversity dependent on those processes can be impacted.[27] Given those concerns, an ideal approach may be to identity conservation targets based on biological diversity and then consider ecological processes.[28] If planning is to focus on processes, one of the challenges is that those processes often operate across a larger area than the typical planning unit.[29] For example, whether the natural flooding of a river continues to occur may be determined far upstream of the location of the conservation project.

Some tools have been developed to incorporate ecosystem dynamics into planning. For example, the minimum dynamic area or minimum dynamic reserve concept enables planners to assess whether a conservation area is large enough to be resilient for known natural disturbances (e.g., windstorms, wildfires). The measure of success is that elements in the system will be retained over explicit time periods although they may fluctuate in size or location.[30] One reason that a conservation vision based on static ecosystems is not an effective model is impending climate change. Future conservation planning efforts will almost certainly find it necessary to incorporate adaptation objectives and actions into their overall strategies to address conservation in the face of climate change.

*Social Issues*

Although scientific or methodological limitations can present barriers to conservation planners, so, too, can social factors. As emphasized throughout, failure to include stakeholders in the planning process early and throughout can doom a project. One of the reasons is that inclusion of stakeholders ensures an informal outreach process. That outreach is especially important today given the increasing disconnect between people and nature.[31] Unfortunately, many scientists are uncomfortable engaging with interested stakeholders, so building bottom-up support for the effort and means of implementation may be limited.[32] Scientists may not be the best leaders of conservation planning processes; conservation practitioners of varied professional backgrounds may be better suited for the task. Success is ultimately dependent on individuals who are involved in and even leading processes to engage with conservation organizations, agencies, and other stakeholders.

The greatest challenge in conservation planning is gaining the public support necessary to implement good conservation plans. In the United States and globally, there is less interest in the environment and conservation than in the recent past.[33] This decreased interest translates to less support both financially and politically. Given this context, it is all the more important that we wisely select areas to invest in conservation and that, at those places, we be strategically smart in addressing critical threats with well-reasoned and practical strategies and actions.

## Conclusion

The biggest hurdles in conservation planning today include the failure of meaningful engagement among scientists, planners, and various public constituencies; the lack of integration from multiple disciplines; and too much focus on the "where" to the detriment of the "how" of conservation planning. Also, planning techniques and approaches must continue to be developed to accommodate the increasing conservation challenges of today. From continuing habitat fragmentation to global warming, the challenges are immense. Despite the challenges, the new paradigm of conservation planning in the twenty-first century—enabling planning across ownership boundaries, engaging local stakeholders, and empowering bottom-up mechanisms—already has shown promise in regions like the Madison Valley of Montana.

**Notes**

1. E. Sanderson et al., "The Human Footprint and the Last of the Wild," *BioScience* 52 (2002): 891–904.
2. J. Hilty et al., *Corridor Ecology: The Science and Practice of Linking Landscapes for Biodiversity Conservation* (Washington, DC: Island Press, 2006).
3. C. Groves, *Drafting a Conservation Blueprint: A Practitioner's Guide to Planning for Biodiversity* (Washington, DC: Island Press, 2003).
4. K. Redford et al., "Mapping the Conservation Landscape," *Conservation Biology* 17 (2003):116–131; K. Chan et al., "Conservation Planning for Ecosystem Services," *PLOS Biology* 4 (2006): 2138–2152.
5. Redford et al., "Mapping the Conservation Landscape."
6. Groves, *Drafting a Conservation Blueprint.*
7. A. Knight et al., "Designing Systematic Conservation Assessments that Promote Effective Implementation: Best Practice from South Africa," *Conservation Biology* 20 (2006): 739–750.
8. Redford et al., "Mapping the Conservation Landscape."
9. Hilty et al., *Corridor Ecology.*
10. P. Landres et al., "Ecological Effects of Administrative Boundaries," in *Stewardship Across Boundaries*, ed. R. Knight and P. Landres (Washington, DC: Island Press, 1998), pp. 39–64.

11. Hilty et al., *Corridor Ecology.*
12. R. Gardner et al., "Neutral Models for the Analysis of Broad-Scale Landscape Pattern," *Landscape Ecology* 1 (1987): 19–28.
13. Hilty et al., *Corridor Ecology.*
14. C. Chester, ed., *Conservation across Borders: Biodiversity in an Interdependent World* (Washington, DC: Island Press, 2006).
15. B. Cestero, *Beyond the Hundredth Meeting: A Field Guide to Collaborative Conservation on the West's Public Lands* (Tucson, AZ: Sonoran Institute, 1999).
16. Hilty et al., *Corridor Ecology.*
17. Groves, *Drafting a Conservation Blueprint*; C. Groves, "The Conservation Biologist's Toolbox for Planners: Advances, Challenges, and Opportunities," *Landscape Journal* 27 (2008): 82–96.
18. E. Sanderson et al., "A Conceptual Model for Conservation Planning Based on Landscape Species Requirements," *Landscape and Urban Planning* 58 (2002): 41–56; The Nature Conservancy, *Conservation Action Planning Handbook* (Arlington, VA: Author, 2007), www.conservationgateway.org/cap.
19. R. J. Lambeck, "Focal Species: A Multi-Species Umbrella for Nature Conservation," *Conservation Biology* 11 (1997): 849–856; T. Caro and G. O'Doherty, "On the Use of Surrogate Species in Conservation Biology," *Conservation Biology* 13 (1999): 805–814.
20. W. Lidicker and W. Koenig, "Responses of Terrestrial Vertebrates to Habitat Edges and Corridors," in *Metapopulations and Wildlife Conservation*, ed. D. McCullough (Washington, DC: Island Press, 1996), pp. 85–109.
21. D. Newmark, "A Land-Bridge Island Perspective on Mammalian Extinctions in Western North American Parks," *Nature* 325 (1987): 430–432.
22. Groves, "The Conservation Biologist's Toolbox."
23. A. Balmford and R. Cowling, "Fusion or Failure? The Future of Conservation Biology," *Conservation Biology* 20 (2006): 692–695; Groves, "The Conservation Biologist's Toolbox."
24. Landres et al., "Ecological Effects of Administrative Boundaries."
25. Groves, *Drafting a Conservation Blueprint.*
26. Ibid.; Balmford and Cowling, "Fusion or Failure?"
27. Lidicker and Koenig, "Responses of Terrestrial Vertebrates."

28. Groves, *Drafting a Conservation Blueprint.*

29. D. Theobald et al., "Ecological Support for Rural Land-Use Planning," *Ecological Applications* 15 (2005): 1906–1914.

30. S. Leroux, F. K. A. Schmiegelow, R. B. Lessard, and S. G. Cumming, "Minimum Dynamic Reserves: A Framework for Determining Reserve Size in Ecosystems Structured by Large Disturbances," *Biological Conservation* 138 (2007): 464–473.

31. O. Pergams and P. A. Zaradic, "Is Love of Nature in the U.S. Becoming Love of Electronic Media? 16-Year Downtrend in National Park Visits Explained by Watching Movies, Playing Video Games, Internet Use, and Oil Prices," *Journal of Environmental Management* 80 (2006): 387–393.

32. Groves, "The Conservation Biologist's Toolbox"; D. Murphy and B. Noon, "The Role of Scientists in Conservation Planning on Private Lands," *Conservation Biology* 21 (2007): 25–28.

33. Pergams and Zaradic, "Is Love of Nature in the U.S. Becoming Love of Electronic Media?"

CHAPTER SIX

# COMMUNITY PLANNING: CHALLENGES, OBSTACLES, AND OPPORTUNITIES

*Lora A. Lucero*

As I walked up to the entrance of the community center in this small northern New Mexico village, I was greeted by an old-timer who asked if I was carrying a gun. This was in 1994, years before 9/11 and the subsequent security checks that occur everywhere today. I was shocked, not sure at first whether he was serious or pulling my leg. However, I quickly realized that he was semiserious because he expected sparks to fly in the meeting I was about to facilitate. I'd been invited to assist the county and this rural village in preparing its first-ever comprehensive land-use plan. My goal for this meeting was to learn what the residents of the community desired for the future of their village.

Without a word spoken, I saw the tension in the packed room. The old-timers, predominantly Hispanic farmers whose families have lived in the village for generations, were standing along three walls, most with their arms crossed against their chests. I soon learned that the younger people sitting cross-legged on the floor in the center of the room were recent newcomers from out of state who were building a commune within the village. Neither group acknowledged the other.

After some "break-the-ice" discussion about the land-use planning process, I asked the old-timers first, "What do you hope to see here in the future?" They took turns describing the past and how their families have lived on the land for generations. They felt a strong connection and respect for the land and didn't want to see it harmed. Most of them lived on ranches they inherited from their families, and when *abuelita* got too old to live on her own, they hoped to build a casita next to the house for her. They also hoped that their daughters and sons who had left to attend college or get jobs in the big city might eventually want to move home. When that day came, they planned to build another adobe on the other side of the property for their grown children. They wanted their families close, the water to flow in the acequias, neighbors who know one another and lend a helping hand when needed, and a healthy landscape to sustain their families.

Then I turned to the young people sitting on the floor and asked them the same question, "What do you hope to see here in the future?" They explained their dream of building a small sustainable community for themselves and their extended friends and family. They hoped to share meals and living space together in a central building, with smaller casitas nearby. They wanted to work the earth and grow their own food, hoping to live off the grid as much as possible. They wanted their human footprint to be as gentle as possible on the land.

Watching the body language and the faces of everyone in the room, I could tell immediately that a huge lightbulb had turned on, figuratively speaking, for nearly everyone in the room. Perhaps for the first time, the old-timers learned what those young "troublemakers" were up to, and the newcomers heard the wisdom and experience that the old-timers had to share. They learned that they were not too far apart from each other. The similarities between their stories were striking.

* * * * * *

Lightbulbs are turning on among conservationists, land-use planners, policy makers, and members of the general public, too, as we begin to appreciate the myriad disconnects that undermine our goals for a sustainable future. This chapter describes some of those disconnects and makes recommendations for mending them.

## The Challenges

The ultimate conservation issues of our time are human population growth and our need to create sustainable human communities while reserving space for other creatures. We need to embrace the challenge of building sustainable communities—including new ways of thinking about the challenge—if for no other reason than to protect the polar bears and all other creatures and their habitats.

The numbers tell the story. More than 6 billion people call Planet Earth home, and if present trends continue, we will add half that many in a few short years—reaching 9 billion by 2050. At the dawn of the twenty-first century the U.S. population was about 282 million, and we can expect to make room for almost 420 million humans by 2050.[1] What impact is this growth, and its concomitant increase in consumption, having on our planet and other species?

By all indications, many of our cities and towns are becoming less desirable places to live and less "people-friendly." Few people walk anymore because we have built roadways to serve cars, not people. Many children have never had the pleasure of walking to their neighborhood school with their classmates, not because they wouldn't want to but because it's nearly impossible to make it through the maze of cul-de-sacs and gated communities. Meeting friends for shopping or a movie requires a car trip; bicycling is for the adventurous. Communities with reliable, safe, and convenient public transportation are the envy of everyone.

The fumes and cacophony of 247 million registered vehicles dominated the landscape in 2005,[2] with ribbons of roads and highways enabling the never-ending sprawl of ticky-tacky subdivisions. As Pete Seeger once sang,

*Little boxes on the hillside, little boxes made of ticky tacky*
*Little boxes on the hillside, little boxes all the same.*
*There's a green one and a pink one and a blue one and a yellow one*
*And they're all made out of ticky tacky and they all look just the same.*[3]

Climate change may very well be the wake-up call for real change. There is little doubt that humans are impacting the climate in serious ways with our dependence on fossil fuel and our profligate consumption patterns. The numbers are going off the charts—rising global temperatures, atmospheric carbon dioxide, and greenhouse gas emissions.

Rajendra Pachauri, a scientist and economist who heads the United Nations' Intergovernmental Panel on Climate Change (IPCC), said in November 2007, "What we do in the next two to three years will determine our future. This is the defining moment."[4] The IPCC has sounded the wake-up call and concludes that "making development more sustainable can enhance mitigative and adaptive capacities, reduce emissions and reduce vulnerability, but there may be barriers to implementation."[5] Those same IPCC scientists predict that the "resilience of many ecosystems is likely to be exceeded this century by an unprecedented combination of climate change, associated disturbances (e.g., flooding, drought, wildfire, insects, ocean acidification), and other global change drivers (e.g., land use change, pollution, over-exploitation of resources)."[6]

Turning to the natural world, the numbers appear just as bleak. The recent rapid loss of species is estimated by some experts to be between one hundred and one thousand times higher than the expected natural extinction rate.[7] According to the World Conservation Union, "Unlike the mass-extinction events of geological history, the current extinction phenomenon is one for which a single species—ours—appears to be almost wholly responsible."[8] More than 15,500 species are known to be threatened with extinction. As humans encroach into the wild and natural places with our sprawling urban-suburban-exurban communities, we gobble up ecosystems, in most cases irrevocably altering the landscape, to the detriment of nature's critters.

The challenges facing the next generation of conservationists—facing us all—may seem daunting. We may be tempted to bury our heads in the sand or curl up with a good book, hoping for the best, but no such luck!

**The Obstacles (or The Big D)**

How did this happen? Certainly no land-use plan ever envisioned this future for its community. No city councilor, planning commissioner, developer, conservation ecologist, urban planner, or community activist knowingly contributed to this predicament—a world on an unsustainable path. Yet each of us has unwittingly done our part to create the mess.

We could turn to many culprits—political, legal, and social—to find part of the answer, but the biggest culprit is what I call "the big D": the many *disconnects* between man and nature. The farmers in that northern New Mexico village believed themselves to be connected to the land, and so did the newcomers; but both felt disconnected from the other. The big D begins in our minds and then finds itself insinuated into our laws and policies, our decision-making and governance models, even our habits and daily routines.

The human brain appears to be hard-wired to see and understand the world in terms of our own worldview, rather than through connections and commonalities.

Picture a silo—that metallic-colored cylinder rising high above the landscape, storing wheat or corn perhaps and designed to protect what's inside and keep what's outside out. The way we live and think resembles the way such a silo functions. We have the ability to separate, divide, and categorize our world in many different ways. We are stuck in silos of our own making—each one undermining any meaningful conservation efforts.

Silo-thinking might have served us well in an earlier age with different challenges, but today such thinking is obsolete and, worse, threatens our very survival.

Silo-thinking comes in many different forms, but four that negatively affect successful conservation efforts include the following:

- *Silo 1: Humans inside the silo, nature outside.* In Silo 1 we find a mind-set in which humans are the dominant species and all others take a backseat, if they are acknowledged at all. The natural world and the services it provides are not well understood or appreciated, and are perhaps even derided by some as existing merely for our entertainment. When push comes to shove, human needs and wants will prevail, and woe to the silvery minnow that may not have enough water in the river to survive, or any other species or habitats that might stand in the way of "progress" and those ticky-tacky subdivisions. Thom Hartmann described this silo when he discussed today's "younger culture"—those who believe humans are not an integral part of, but are separate from, the world in which they live.[9] In this silo it is human beings' destiny to subdue and rule the rest of creation outside of the silo.
- *Silo 2: Nature inside the silo, humans outside.* In Silo 2 we find the opposite. In this silo, the natural landscape, pristine and untouched by human hands, along with the innumerable critters that call it home, are wrapped up in a protective cocoon, while people and their sullied towns and cities are banished. The mind-set represented in Silo 2 prefers to either ignore the interaction and connections between humans and nature or, worse yet, actively impede such connections. Some might argue that this silo represents balance and justice for the natural world, but it is merely a mirror image of the mind-set we find in Silo 1.

- *Silo 3: My city inside the silo, the surrounding region outside.* In Silo 3 we find "local control" and the parochialism inherent in local decision making that often contravenes the broader needs of the region. Growth and development are typically local issues; most states have delegated their land-use and planning authority to local governments. At the same time, cities are very competitive creatures, and their need to enhance their tax base plays a large role in that competition. So while the people of the region need clean air and a safe and reliable transportation system, unless the state retains some of its land-use and planning authority, competing municipalities will continue to grow and develop without regard to the impacts their decisions may have on the region. Although a particular municipality may reap the benefits of the tax receipts from its new Wal-Mart, the resulting traffic snarls and air pollution will not respect municipal limits. The region will suffer the consequences, even though the region has little or nothing to say about that growth.
- *Silo 4: My family and private property inside the silo; my neighbors, the public commons, and the general welfare of the community outside.* Silo 4 is the most pernicious of them all. In Silo 4 private property rights are supreme, the public commons be damned. Imagine a country where the government must pay the property owner *not* to clear-cut all of the trees on his private forestland because he wants to pave it over with a shopping center, or where the government must compensate the property owner for the air rights over his property within the approach zone of an airport. Both examples are happening today. The balance between private property rights and the rights of the public has been skewed so heavily in favor of the first that the rights of the public and nature are scarcely on the table when decisions about growth and development are made.

Each of these silos disconnects us in one way or another from understanding the fundamental truth—that humankind and nature are interconnected—but our laws, regulations, and land-use plans unfortunately reflect and perpetuate the disconnects. Just as Newton's three laws of motion are accepted as undeniable truths of nature, so must we understand and appreciate our connection to land and nature.

**The Opportunities**

Community planning is about "seeing" the future, recognizing the challenges, and designing a blueprint (the plan) for achieving our goals for the future. But community planning and planners today are under fire from all sides. Perhaps they make an easy target. When the results don't match our expectations, where else are we to lay the blame?

The community planning process typically involves the community residents attending many meetings and expressing their hopes and concerns for the future. A plan is eventually prepared and then adopted by the elected officials, and everyone pats each other on the back and goes home, thinking their community has been moved one step toward a better future. Unfortunately, in many cases it hasn't.

Many obstacles conspire to thwart the community's successful implementation of its land-use plan today. In the vast majority of states in this country, the laws that govern land-use planning are based on the Standard State Zoning Enabling Act (SZEA), a model enabling statute drafted in 1922, and the Standard City Planning Enabling Act (SCPEA), drafted in 1928.

Imagine the challenges confronting those early drafters nearly a century ago, and imagine how different the challenges are facing our communities today. We have more than doubled our population; we drive Hummers and hybrids, not the Model T Ford; we build superstores of more than 250,000 square feet that offer nearly any item imaginable for sale, not the small neighborhood shops along Main Street; and many people are meeting their future mates online, not at the ice-cream social. Government was simpler in the 1920s, and there were fewer governmental units. Today, the number of local governments has grown by leaps and bounds, and each has some jurisdiction for land-use planning.

Back in the 1920s, we had a different regard for land and how we used the land; we were more likely to view land as a mere commodity to be bought and sold, rather than a resource with many intrinsic values, such as the importance of wetlands for flood control and wildlife habitat. Today, people are more actively engaged in community planning. In the 1920s, the focus was on protecting the public from nuisances—keeping the pig out of the parlor.

Today, planners must certainly address nuisances, but they must also anticipate the needs of a burgeoning population, the dwindling fiscal and natural resources, interjurisdictional conflicts, private property rights, and climate change—to name but a few. The SZEA and SCPEA—the legal basis for much of our land-use planning today—could not have anticipated these changes. Some states have begun to update and reform their planning and land-use enabling laws, but progress has been slow. Entrenched development interests that reap the benefits of the status quo are reluctant to support changes unless they enhance private property rights (e.g., Silo 4).

Not only are state planning laws antiquated, but most states fail to require communities to follow the land-use plans they have adopted. Those plans are merely "advisory," to serve as guides for decision makers. When push comes to shove, nine times out of ten the plan gets shoved aside. We are, in fact, entrusting the private sector to fulfill the community's goals for the future by allowing developers, in many cases, to dictate where, when, and how the community will grow. Most land-use development decisions are made at the request of a developer or property owner, with the local government reacting to ad hoc proposals, not proactively engaged in creating the community.

Furthermore, the land-use plan may call on decision makers to adopt a whole host of regulatory tools and incentives to accomplish the community's goals. The most typical tools are zoning ordinances, subdivision regulations, impact fee ordinances, and capital improvement programs. If the tools are not in sync with the land-use plan—not consistent with the adopted goals and policies—it is difficult to accomplish the plan's goals. Unbelievably, the majority of states still do not require that such land-use tools be consistent with the adopted land-use plan.

Community planning has become conflated in the minds of many with land-use development. Land-use planners in some communities have been conscripted to be development permitters, while politics and the politicians have greater influence over the ultimate use of land. The public is generally unaware that there is a major disconnect between the plans they help to create and the day-to-day implementation of those plans. Perhaps we have identified another silo.

What is it going to take to fix the big D and bust the silo-thinking? Is population growth on a collision course with a planet stretched to the limits? Are the disconnects so prevalent and entrenched that they cannot be mended? We have examples of how humans have taken a very different path when confronted by big challenges. Consider the agricultural and industrial revolutions. We devised new methods for energy production, food production, and manufacturing when it became necessary. We need a silo-busting revolution today that equals those earlier revolutionary efforts.

Climate change will no doubt be the catalyst for the present generation to choose a different path. If we are successful in meeting that challenge, we will all be conservationists to the core—truly appreciating and understanding our connection to ecosystem services. We will be judging our actions today based on how they might affect the seventh generation in the future,[10] rather than on how they might be reflected in the next Gallup Poll or the gross domestic product.

Community planning is merely a tool, not the solution, but it is very important. The single biggest fix we can achieve is to reform the antiquated state laws that govern land-use planning and statutorily require consistency between plans, regulations, and decision making. Implementation of land-use plans must be more closely aligned with our intentions. The incremental, day-to-day decisions must be tied to the community's comprehensive plan, rather than to what seems to be politically expedient in the short term. This will also provide greater accountability; the public will be able to measure progress—or lack of progress—in implementing the community's plan.

Relatively new land-use regulatory measures have been introduced into the community's toolbox (table 6.1), but they are only as good as the plan the implement and the ecological information on which they are based. The growing body of science and ecological information available today will certainly become more accessible to decision makers, planners, and the public tomorrow.

| **Table 6.1**<br>**Land-Use Regulatory Measures for Conservation** |
|---|
| Transfers of development rights (TDRs) and purchases of development rights (PDRs) to channel new development away from areas of the community that should be protected and into areas where higher-density development is appropriate. |
| Urban service areas (USAs) and urban growth boundaries (UGBs) to demarcate where new development should and should not go. |
| Conservation overlay zoning and cluster subdivision regulations to minimize the footprint of new parcels on the landscape and to protect natural areas such as wetlands. |
| Targeted protection standards—e.g., stream and wetland buffers, tree and vegetation protection, hazardous area protection. |
| Conservation easements to preserve natural areas and agricultural and forest lands. |

Land-use planning tomorrow is going to begin with a new question. Rather than asking ourselves where the best locations for development are in our communities, we will be asking, "Where are nature's vital and critical locations in this region that must be preserved and protected?" Today we have our priorities backward. We need to be asking the right questions in the right order.

Community planning tomorrow will reexamine and reprioritize our assumptions about costs and benefits. When we factor in the costs and benefits of development to the community today, we rarely include the costs associated with habitat loss, for example, or the benefits of habitat restoration. Nature has many real and intrinsic benefits that must be included in the calculus of any fiscal impact analysis. As that information becomes more apparent to the public, a parade will form demanding greater care and protection for all of nature's species and their critical habitats. Decision makers will be scrambling to rush to the front of the parade.

There will be no land-use plans sitting on the shelf gathering dust. Community planning tomorrow will be a complete loop, with the public involved each step of the way. First, the plan will be based on a stronger foundation of information. After the plan is adopted, it will be actively monitored to see whether the incremental steps taken to implement it are consistent with the plan. But we won't stop there. We will evaluate the impacts of our decisions to determine whether the plan needs to be adjusted. No plan will ever be cast in concrete.

There is going to be a seismic shift in how the public views land-use planning in the future. Today, land-use planning equals land-use development. Undeveloped land is somehow underutilized, is inferior, and should be put to its "highest and best use." We are going to rethink our relationship to the land. Land conservation will be elevated in importance because the public, policy makers, and decision makers will understand the connections between humans and nature as the silo-thinking disintegrates.

Land conservation does not mean that growth and development will be stopped in its tracks, but the long-term dimension will take on greater significance in our "short-term" decision making. We will, we hope, make wiser use of the land and appreciate the many potential uses—including nature's needs—when we make development decisions. Rebalancing the rights of the private property owner with the rights of the public and nature may seem unthinkable in the current political milieu, but land-use planning in the future will play a role in that rebalancing. The absurdity of carving up the hillside so that a 6,000-square-foot house can be built on a 30 percent or greater slope or paving over a wetlands for a superstore will be understood for what it is—absurd! Our children's children's children will look back and say, "Thank goodness our great-grandparents chose a different path."

Community planning tomorrow will occur on a silo-free landscape when we begin to mend the disconnects that keep humans and nature apart. Our plans today may make us feel secure about the future, reflecting our sense that we can control nature and the landscapes where we live, but tomorrow we will have greater humility.

## Conclusion

Conserving nature and ecosystem services requires an active and intentional plan to build sustainable habitats for people, because our prevalent sprawling settlement patterns pose significant risks to the survival of all species. We should change the rules of the development game to foster more sustainable communities by connecting land-use planning and the development process—beginning with the way we think about our relationship to nature. When community planning connects the visions of both the longtime farmers and the recent newcomers, and the land-use laws and regulatory framework enable their visions to be implemented, the silos will come crashing down.

**Notes**


1. U.S. Census Bureau, 2004, "U.S. Interim Projections by Age, Sex, Race, and Hispanic Origin," http://www.census.gov/ipc/www/usinterimproj/.
2. U.S. Department of Transportation, Bureau of Transportation Statistics, http://www.bts.gov/publications/national_transportation_statistics/html/table_01_11.html.
3. "Little Boxes," words and music by Malvina Reynolds; copyright 1962 Schroder Music Company, renewed 1990, performed by Pete Seeger.
4. Elisabeth Rosenthal, "U.N. Chief Seeks More Climate Change Leadership," *New York Times*, November 18, 2007, http://www.nytimes.com/2007/11/18/science/earth/18climatenew.html?_r=1&ref=todayspaper&oref=slogin.
5. Intergovernmental Panel on Climate Change, "Summary for Policymakers," in *Climate Change 2007: Impacts, Adaptation and Vulnerability, Contribution of Working Group II to the Fourth Assessment Report of the Intergovernmental Panel on Climate Change*, ed. M. Parry et al. (Cambridge: Cambridge University Press, 2007), p. 18.
6. Ibid., p. 11.
7. World Conservation Union, Species Extinction, http://iucnredlist.org/.
8. Ibid.
9. T. Hartmann, *The Last Hours of Ancient Sunlight: The Fate of the World and What We Can Do Before It's Too Late* (New York: Three Rivers Press, 2000).
10. The Great Law of Peace of the Haudenosaunee (Six Nations Iroquois Confederacy), http://www.degiyagoh.net/worldview_haudenosaunee.htm.

CHAPTER SEVEN

# ECONOMIC INCENTIVES: CONSERVATION THAT PAYS

*Luther Propst, Adam Davis, John Shepard, and Nina Chambers*

Aldo Leopold once wrote that "the oldest task in human history is to live on a piece of land without spoiling it." The Sonoran Institute was founded on the belief that the challenge Leopold laid out over seventy years ago is as relevant now as it was then. Given the increasing rate of development of open lands, and the parallel increases in the value of real estate, it would be prohibitive for anyone to purchase all of the important private conservation lands. The Sonoran Institute, along with an increasing number of private conservation organizations, believes there has to be a way to work with land developers to integrate conservation into their projects and to have the development process pay for conservation.

The Sonoran Institute came up with the concept of community stewardship organizations (CSOs), conservation organizations created as part of development projects that would manage permanently protected open space within those projects. To underwrite those new organizations, the developer agrees to a series of perpetual funding mechanisms, including real estate transfer fees, tied to the development project.

CSOs have now been established across the United States, protecting more than 20,000 acres of private land. Their reach extends beyond development projects with which they are affiliated, as they now hold easements on nearby lands, undertake ecological restoration projects, and organize environment education programs for neighborhood schools.

This chapter explores the economies of a new way of doing conservation. This new approach sees conservationists partnering with landowners in large-scale habitat conservation programs, as well as exploring innovative ways of funding those programs. The immediate challenge is to convince a broader section of the development community and the financial institutions that serve it that integrating conservation into land development pays financially by reducing the cost of complying with environmental regulations, by attracting a larger share of the home buying market, and because real estate in those developments appreciates at a faster rate than conventional development projects.

* * * * * *

The financial resources in the United States that are targeted for biodiversity conservation, ecosystem services such as clean air and water, and protection of landscape character are insufficient to realize those goals. While public agencies, conservation organizations, land trusts, and conservation-oriented private landowners continue their good works, the pace of urban expansion, resort and rural housing construction, highway building, energy development, and natural resource extraction is unsustainable.

The pace and scale of land development, the high price of land, and the perennial challenge of underwriting the cost of monitoring and managing conservation lands indicate that while direct acquisition of individual conservation properties is desired and necessary, it alone is an insufficient conservation strategy.

Other conservation approaches warrant increased investigation and investment. One that is highly promising is to invest in innovative market mechanisms and incentive programs that reward private landowners—and public land managers—for each scientifically measurable unit of environmental improvement on their property. Targeted investment in that kind of structural solution could improve conservation and land-use decisions across millions of acres.

## Background

Designating or withdrawing public lands for conservation purposes was the first method of protecting land with high scenic or conservation value. Much later, many conservation organizations, led by The Nature Conservancy, pioneered the acquisition of private lands for conservation with charitable funds. Land protection funders and professionals soon realized that that approach would prove inadequate. Those groups then started investing in campaigns to win approval of local and state bond measures and other funding mechanisms for acquiring conservation lands. Awareness developed gradually of just how significant that more leveraged approach would be. Local finance measures were pioneered in college towns (e.g., Boulder, Colorado, in 1967) and resort communities (e.g., Nantucket, Martha's Vineyard, and Block Island in the 1980s) with support from the conservation community. State bond measures developed in California and New Jersey in the 1970s were followed in the 1980s by "Environmental Endangered Lands" bonds promoted by The Nature Conservancy in Florida. In the mid-1990s that conservation tool went to scale as the number of conservation bond measures across the country exploded when the Trust for Public Land launched its Conservation Finance Program and communities everywhere began using local land protection to grapple with poorly planned growth. The Trust for Public Land and the Land Trust Alliance publicized that explosion through LandVote research and publications. LandVote is a database source of information maintained by the Trust for Public Lands to facilitate research on conservation bond measures.

## An Emerging Strategy

Market mechanisms and financial incentives for conservation represent another major strategy to leverage the effectiveness of conservation investments. Such tools may be at a stage similar to that of state and local finance campaigns fifteen or twenty years ago.

Evidence shows that market mechanisms and financial incentives work to conserve land and promote restoration. The global climate market, for example, saw an increase in growth from an approximate value of $12 billion in 2005 to $29 billion in 2006.[1] The voluntary carbon market has approximately doubled its volume every year, with 2004 at three million to five million tons per year and 2006 at an estimated twenty million to fifty million tons per year.[2]

Conservation banks have also seen huge increases over the last decade; just a few existed in 1995, and by 2003 there were more than thirty.[3] Conservation banking's annual estimated turnover is between $200 million and $300 million.[4] Wetlands mitigation banking saw a 376 percent increase in approved banks—with 110 banks in 1992 and 526 banks in 2005. The number of banks that have sold out of credits has tripled since 2001 due to increasing demand. Another measure of success is that wetland banks were mostly in Florida and California in 1992, but they have since been established in twenty-nine states.[5] The latest figures from *Ecosystem Marketplace* indicate that wetland mitigation in the United States is worth $1 billion per year.[6]

Over the past twenty years, regulatory mechanisms have been developed to reward incremental provision of ecosystem services by private landowners. These mechanisms are relatively new but have a clear track record of success in aligning environmental and economic outcomes. Given the early success and promise of approaches that reward conservation on private lands, additional support and development are warranted.

## Need and Opportunities

Three critical factors should be considered if market mechanisms are to be applied at a scale that is meaningful to conservation policy:

1. The loss of natural landscapes and the environmental services they provide has progressed to a point where scarcity *in relation to human needs and human numbers* is increasing the financial value of those landscapes and services. In some cases, environmental services are now sold or traded through emerging markets. In other cases, they are not, but supply and demand of environmental services are now important factors affecting land-use and land transactions.
2. Regulatory-driven market mechanisms and incentive programs usually emerge to address one set of environmental services, generally operating in "silos" where the cumulative benefit of conservation actions is not acknowledged or financially rewarded.
3. Environmental goals and policies are inconsistently applied across a wide range of landownerships, including private *and* public lands.

Between 2000 and 2030, the United States is projected to grow by more than 82 million people, an increase of more than 29 percent.[7] Three states would account for nearly half of the population increase: Florida, California, and Texas. The five states with the greatest projected growth rate are Nevada, Arizona, Florida, Texas, and Utah. Most of the population growth is projected to occur in the South and the West. The share of the U.S. population living in those regions would increase from 58 percent in 2000 to 65 percent by 2030.[8]

As the country grows in population, it is likely to accommodate much of the new growth in previously undeveloped areas ("open space"). Developed lands already outnumber protected private lands. Given projected population growth and the likelihood that "infill" development will provide no more than one-third of new housing,[9] there is tremendous potential for development to vastly outpace conservation efforts.

Land development, while it is the most pressing trend, is only part of the challenge facing private land conservation efforts. Climate change, invasive exotic plant species, fire suppression, poor grazing practices, logging, groundwater pumping, surface water diversion, erosion, and other impacts are severely degrading natural landscapes on both private and public lands, including areas that are officially protected.

Finally, certain critical habitat types are being converted at particularly high rates. Notable examples are ponderosa pine, hemlock-sitka, and lodgepole pine forest cover types, which together have decreased by at least 23 million acres since 1963.[10] Wetlands, riparian areas, and watercourses are especially threatened, particularly in the arid and semiarid Colorado River Basin. A recent assessment of the threat to wetland and riparian area plant communities showed that 12 percent of the nation's 1,560 wetland communities are critically imperiled, 24 percent are imperiled, and 25 percent are vulnerable.

It is ultimately this unraveling of natural landscapes that matters. Acres lost to development serve merely as a proxy for the larger issue: the loss of biodiversity, water, air quality, recreation, and other related ecosystem services. Essentially, the growing relative scarcity of natural landscapes and ecosystem services has progressed to the point where specific actions protecting or restoring them can provide revenue from multiple sources. Such funding sources now include the following:

- "Markets" created by mitigation requirements in the Clean Water Act and the Endangered Species Act
- Consumer demand for development that integrates conservation of natural areas into the process
- "Conventional" easement sales and public payment schemes such as those in the conservation title of the Farm Bill

Worldwide, there is increasing recognition of the value of ecosystems because the water, biodiversity, and climate regulation services they provide are no longer enormously abundant in comparison to human uses and human needs.[11] While regulation, land-use planning, and environmental laws all address this situation to some extent, it has become clear that additional approaches are needed. While traditional regulation can stop the worst behaviors, it seldom brings out the best management.

**Developing Environmental Markets**

In response to changing conditions and the limitations of existing approaches, a range of market interventions has developed over the past two decades (see table 7.1 for examples), directly or indirectly enabled by underlying regulation. These include mitigation banking, conservation banking, water quality trading, and carbon offsets, along with integrated conservation-development approaches that include limited or conservation development, sustainable timber and agriculture, and tradable development rights.

In the case of hybrid programs, such as limited or conservation development or certified timber production, the consumer demand for the environmental benefit is in some sense subsumed into a more traditional commodity (e.g., housing or timber). For example, according to a detailed statistical analysis in a 2006 article in *Urban Affairs Review*, "Conservation subdivisions carry a premium, are less expensive to build, and sell more quickly than lots in conventional subdivisions."[12] The market for certified timber, which began in 1993 when Forest Stewardship Council standards were developed, has resulted in improved management practices on over 200 million acres of timberland worldwide.[13] In these cases, there is no formal mechanism, but the value to the buyer of the house is bolstered by the environmental benefits associated with the products.

A wide range of economic activities are now required to "offset" certain environmental impacts. For example, from local land-use planning goals spring tradable development credits and density credits; impaired water quality leads to water quality trading; unavoidable "take" of endangered

**Table 7.1**

**Examples of Conservation Results through Financial Incentives**

| Name of Program | Location | Summary of Results |
| --- | --- | --- |
| New York City water system | Croton Watershed Region | One of the most successful nonpoint pollution control programs in the United States, it has played a major role in stabilizing and reducing watershed pollution loads and in enabling the city to avoid the multi-billion-dollar cost of filtering its water supply. Within five years after this program was established, 93 percent of all farmers in the watershed had chosen to participate. |
| PowerTree Carbon Program | Lower Mississippi Alluvial Valley in Southeastern Arkansas | In 2003, 25 companies established the PowerTree Carbon Company, LLC, a voluntary consortium of leading U.S. electric power companies that have committed $3 million to establish six bottomland hardwood reforestation projects in the Lower Mississippi Alluvial Valley states of Louisiana, Mississippi, and Arkansas, with the federal government, conservation groups, and landowners as partners. As the trees grow, they will eventually capture more than 1.6 million tons of carbon dioxide from the atmosphere—as well as provide critical habitat. |

| | | |
|---|---|---|
| Ecological Enhancement Program | North Carolina | In its first two years of operations, the initiative facilitated more than $1.9 billion in road construction across the state without a single project delay because of a lack of mitigation. Along with managing nearly 400 stream and wetland restoration projects, the Ecological Enhancement Program collaborated on the preservation of nearly 35,000 acres of natural areas for future generations. In 2005, the program was designated as one of the top fifty new innovative government programs in the nation by the Harvard University Kennedy School of Government. |
| Conservation banking | California | Over the past decade, California has permanently conserved over 40,000 acres of habitat in conservation banks. Conservation banking allows mitigation to be done on fewer, larger sites than conventional "on site" projects; it allows flexibility to establish banks on sites that may result in greater ecological benefit; and it has reversed the common perception of endangered species on private property from liability to asset. Conservation banking creates the potential for landowners to profit from the conservation of endangered species on their property. |

**Table 7.1 (continued)**
**Examples of Conservation Results through Financial Incentives**

| Name of Program | Location | Summary of Results |
|---|---|---|
| Transferable state tax credits | Colorado, New Mexico, South Carolina, and Virginia | A conservation credit is an income tax credit available to landowners who voluntarily preserve their land through the donation of a conservation easement or fee title. The donation must protect conservation values as defined by individual states and must be made to an entity qualified to hold such property interest by the terms of the legislation creating the credits.[a] The ability to transfer credit to a third party appears to greatly increase the effectiveness of tax credit programs. In Virginia, the average number of donations doubled and the number of acres protected tripled once tax credits were made transferable. Since 2002, an average of 75 percent of credits have been transferred to a third party.[b] |

| | | |
|---|---|---|
| Nontransferable state tax credits | California, Connecticut, Delaware, Georgia, Maryland, Mississippi, North Carolina, and New York | Although the tax credit programs in these states are not transferable, they remain effective tools to protect conservation values on private lands. Besides transferability, the credit cap is another important variable in these laws. In North Carolina, which has the oldest state tax credit program (started in 1983), the average number of donated conservation easements doubled when the credit cap was raised from $25,000 to $100,000 per individual and to $250,000 per corporation. Similarly, North Carolina increased land conservation donations by 20 percent, and New Mexico saw an increase of 25 percent when its caps were raised to $100,000 or above.[c] |

[a]Information taken from Debra Pentz, with Roman Ginzburg and Ruth McMillen, *State Conservation Tax Credits: Impact and Analysis* (Boulder, CO: Conservation Resource Center, 2007), p. 9.

[b]Information taken from Pentz, p. 13.

[c]Information taken from Pentz, pp. 12–14. For more information on state tax credits and a comparison among states, visit the Land Trust Alliance Web site at http://www.lta.org/publicpolicy/state_tax_credits.htm.

species results in conservation banking, which has protected approximately 110,000 acres of endangered species habitat; and impacts to aquatic resources (as defined under section 404 of the Clean Water Act) create mitigation banking. The value of sequestering carbon dioxide through forest management and other land-use change is also beginning to be recognized.

### Market Incentives Limitations

Several challenges or issues are currently limiting the success of implementing market incentives for private land conservation. Below is a discussion of two important issues: (1) regulatory silos that limit conservation benefit to a single function and (2) the need for an integrated approach to both public and private lands to effectively protect ecological function.

#### *Issue 1: Regulatory Silos*

A fundamental challenge in applying these market concepts is that when landowners or land managers are rewarded for only one environmental service (e.g., carbon storage or endangered species protection), they still have an incentive to develop the land because its development value will likely outweigh the value of that single service.

While a set of policies, initiatives, and constituencies address market- and incentive-based approaches to conservation actions, little time and attention are devoted to how those actions can be integrated or made complementary. Revenue from a variety of actions—carbon sequestration, water quality, extinguishing development rights, wetland or stream restoration, water storage, or habitat restoration—may be available individually to a single large parcel owner, but no unifying set of mechanisms exists to reward the landowner for collectively providing these services. There is clearly a need to improve the ability of private landowners to take advantage of the full range of environmentally related values produced by their property.

It is easy to imagine, for example, a set of management actions on a specific ranch resulting in downstream water quality improvements, flood mitigation, wetland restoration, improvement and protection of endangered species habitat, and carbon sequestration. Because of the way requirements are set by the various agencies for each type of outcome, however, the landowner is likely to face trade-offs that reduce the economic value of his investments. If a given acre is "set aside" in a mitigation bank to offset wetland impacts under section 404 of the Clean Water Act, the landowner cannot use that acre for any other revenue-generating purpose, even if that purpose is complementary with wetlands ecological function.

One promising development is happening in Oregon, where the Department of Transportation, in cooperation with the U.S. Army Corps of Engineers, the U.S. Fish and Wildlife Service, and six other state and federal agency partners, is creating a new "currency" that reflects both biodiversity and wetlands values. This new instrument, called a Habitat Value Credit, is designed to recognize that a given acre can provide multiple values and at the same time assure regulators that impacts are fully offset.

*Issue 2: Application to Public as Well as Private Lands*

Consistent with this more integrated approach is a framework that rewards each scientifically measureable unit of environmental improvement on both private and public lands. It is preservation of the land's inherent conservation value, not landownership or function *per se*, that should be the primary consideration. Ownership and function matter in terms of how mechanisms are structured and how performance is rewarded.

Given the extent of public land ownership, market-based regulatory approaches to conservation management should not ignore public lands, including federal land, state trust lands, state parks, and county or municipal land. Specific mechanisms developed to provide incentives to private landowners, such as mitigation banking, conservation banking, water quality trading, and carbon banking, can be used as a basis for enhancing funding for managing and restoring public lands. While U.S. Forest Service lands may never qualify to offset impacts on private lands, measurement of water, carbon, and biodiversity flows against a baseline could be a basis for agency appropriations and a powerful means for articulating the Forest Service mission to the public. Such an approach could also provide the basis for reforming federal payment in lieu of taxes (PILT) appropriations to counties.

In another example, land managed by the U.S. Bureau of Land Management (BLM) in Las Cienegas National Conservation Area southeast of Tucson, Arizona, sources 15 to 20 percent of Tucson's drinking water; a system could be devised to provide budget allocation to the BLM land managers there for each measurable unit of restoration activity they undertake that improves the quality or availability of that water source.

Similarly, such mechanisms can supply the basis for targeting public payment systems such as those within the conservation title of the Farm Bill. The 2002 Farm Bill authorization devotes about $3.5 billion annually to a number of private land conservation programs that provide funding for land management, restoration, conservation easements, water quality protection, and other activities.[14] Tremendous opportunities exist, however, to improve the manner in which payments are targeted to priority areas. The "ecosystem services" framework offers a guiding principle that would reward landowners who invest more, absorb more opportunity cost, and ultimately provide more public benefit from their stewardship.

Attention must also be paid to policies that subsidize ecologically degrading activities or otherwise create incentives that undermine conservation. Government programs that fund road and infrastructure development, facilitate construction of second and third homes through the deductibility of mortgage interest in the tax code, and subsidize insurance for development in flood plains and coastal zones are often cited as leading culprits.

Finally, we should acknowledge the growing evidence documenting the causes and effects of global climate change. Climate change not only threatens to undo much of what has been accomplished with land conservation but may also significantly shift resources in the future that might otherwise be allocated to land conservation. Future conservation efforts must be able to demonstrate how they help to address climate change causes (e.g., reduce energy demand, promote alternative energy) and effects (e.g., preserve and connect habitats vulnerable to climate change, enhance groundwater storage). This connection to climate change underscores the importance of defining, measuring, and communicating the specific outcomes of land conservation action.

**Implementation Strategies**

Increased investment in mechanisms and structural solutions that promote financial incentives may generate significantly higher leverage for conservation management of private land than for direct purchase of lands. The goal is to identify and develop public policy innovations that leverage investments. This is similar to conservation organizations and foundations investing in state and local open-space bonds and other public funding mechanisms that offer significantly higher leverage than direct acquisition.

Effective investment in research, policy development, demonstration projects, and communications and outreach is warranted.

*Research*

Priorities for research include developing or improving ways to quantify the economic value of environmental services and monitoring the extent to which land management choices affect provision of those services. There also needs to be a better understanding and selection of high-level or keystone indicators that can serve as reliable proxies of the economic value of market transactions.

*Policy Development*

To ensure that state and federal policies are complementary, recommendations are needed that outline steps federal and state policy makers can take to integrate and harmonize desired outcomes across the range of potential programs. Federal and state agencies will also need education and outreach to align environmental policy with field implementation. Appropriate funding, fee, and licensing structures will be required to ensure fair and sufficient resources for permitting and enforcement.

*Demonstration Projects*

A diverse portfolio of demonstration projects will show a broad applicability across a range of geographies and contexts. Well-designed pilot programs should demonstrate principles in practice on specific types of private landholdings and then monitor and evaluate those demonstration projects, determining lessons learned and best practices. Priority regions that would make good demonstration projects include the following:

- *Regions that are experiencing rapid urban growth.* These may include California's Central Valley and Sierra Foothills, the Phoenix and Tucson metropolitan areas in Arizona, Utah's Wasatch Front, and Colorado's Front Range.
- *Regions where enhanced private land management would significantly enhance public land values, particularly landscape connectivity and integrity.* Considering the wide-ranging habitat needs of carnivores and ungulates, the corridor between Yellowstone National Park and the central Idaho wilderness complex is one such region.

- *Regions where the economic viability of farms and ranches would be significantly enhanced or where improved water management would benefit salmon restoration through market-based conservation mechanisms.* Such regions might be best defined as water basins or watersheds, such as those encompassing Oregon's Deschutes and John Day rivers.
- *Regions where market influences in water allocation could better serve both wildlife and human communities.* The heavily managed lower basin of the Colorado River is one example.
- *Regions that will likely be affected in particularly adverse ways by climate change.* The Albemarle Peninsula in North Carolina and the Florida coastal zones are examples.

### *Communication and Outreach*

An important part of an implementation strategy is articulating and disseminating the advantages and incentives to the private landowner community, the investment and financial communities, state and federal policy makers, and other partners needed to promote policy reform.

## Obstacles

Numerous technical, logistical, and political obstacles stand between the present and full realization of the potential of ideas introduced here. A major obstacle is the fragmented and diffused nature of government regulation and regulatory agencies (note the federal regulation of pollution by media—air, water, solid waste, toxics). It is important to point out, however, that conservation transactions resulting in both financial reward and clear environmental improvement are already taking place across a variety of media and issue areas, demonstrating that resolution of this structural obstacle is not a requirement for effective progress.

A second set of obstacles is the difficulty in identifying, valuing, monitoring, and communicating the economic and environmental benefits associated with these mechanisms. While relatively clear metrics have been established for preservation and restoration of habitat and aquatic features under mitigation and conservation bank guidance, there remain tremendous variations in how the guidance is interpreted and applied and great inefficiencies in implementation.

Finally, some people on both the political left and the right oppose incentive- and market-based conservation approaches on principle. There are also elements of the regulatory community that see private action for conservation as threatening. It will be essential to continue to overcome objections through the use of effective and accurate communication about those approaches.

**Conclusion**

In sum, there is merit in investing in a new generation of regulatory and market mechanisms that raise incentive-based conservation approaches to a significantly higher level of impact and effectiveness for private lands. To do this effectively a number of considerations need to be addressed.

First, the use of incentive-based mechanisms should be integrated on specific landscapes and large parcels. Optimal demonstration projects sites should be chosen to maximize lessons learned and test the tools. Such an approach would help avoid silos and instead integrate multiple ecosystem values for enhanced benefits for conservation and the landowner.

Second, market-based approaches should be applied to various types of landownership—public and private. Financial incentives can be used for private landowners as well as state land agencies, federal land management agencies, and local or county lands. Application across landownership patterns would also encourage an ecosystem approach to protection across a larger landscape.

Third, regulatory and nonprofit communities should be engaged to advance the transaction efficiency and the scientific measurement efficiency needed. Diverse groups should be engaged in the demonstration sites and actively coordinate efforts for effectiveness.

Last, market incentives for land conservation need to be used at a scale that ensures that conservation policies will be effectively carried out.

**Notes**

1. K. Røine and H. Hasselknippe, eds., *Carbon 2007: A New Climate for Carbon Trading*, Carbon Market Insights 2007, Point Carbon's 4th annual conference, Copenhagen, March 13-15, 2007.
2. M. Kebner, paper presented at Green T Forum: Raising the Bar for Voluntary Environmental Credit Markets, New York, May 1–2, 2006.
3. J. Fox and A. Nino-Murcia, "Status of Species Conservation Banking in the United States," *Conservation Biology 19,* (2005): 996–1007, no. 4
4. R. Bayon, managing editor of *Ecosystem Marketplace*, personal communication, San Francisco, 2007.
5. J. Bishop, S. Kapila, F. Hicks, and P. Mitchell, *Building Biodiversity Business: Report of a Scoping Study*, discussion draft (London and Gland, Switzerland: Shell International Limited and World Conservation Union, 2006), p. 58.
6. *Ecosystem Marketplace*, http://ecosystemmarketplace.com/.
7. R. Bernstein. "Florida, California, and Texas to Dominate Future Population Growth, Census Bureau Reports." U.S. Census Bureau News. www.census.gov/Press-Release/www/releases/archives/population/004704.html.
8. Ibid.
9. D. Priest, "Planned Communities and the Smart Growth Movement," working paper cited in Heid and Greenfield, eds., *Development without Sprawl: The Role of Planned Communities* (Washington, DC: Urban Land Institute, 2004).
10. Environmental Protection Agency's draft of *Report of the Environment 2003*, Technical Document, pp. 5–12.
11. For comprehensive evidence of this, the reader is referred to the Millennium Ecosystem Assessment, the largest and most comprehensive study ever undertaken of the health of the world's ecosystems; prepared by 1,360 experts from ninety-five countries with extensive peer review, it represents a consensus of the world's scientists. Available online at www.millenniumassessment.org.

12. R. Mohamed, the Economics of Conservation Subdivisions: Price Premiums, Improvement Costs, and Absorption Rates," *Urban Affairs Review,* vol. 41 (January 2006): 367–99.
13. Forest Stewardship Council, www.fsc.org/en/about/about_fsc.
14. This discussion of Farm Bill incentives and opportunities is taken from "Financing Private Lands: Conservation and Management through Conservation Incentives in the Farm Bill," chapter 12 in *Walden to Wall Street*, ed. J. Levitt (Washington, DC: Island Press, 2005).

CHAPTER EIGHT

# ECOSYSTEM SERVICES: THE NATURE OF VALUING NATURE

*J. B. Ruhl*

In the spring of 1998, Gretchen Daily of the Stanford Biology Department, Geoff Heal of the Columbia Business School, and Peter Raven, director of the Missouri Botanical Gardens in St. Louis, organized a conference in St. Louis to explore the concept of ecosystem services, a relatively new idea within the disciplines of ecology and economics. I had read Daily's groundbreaking edited volume on the topic, *Nature's Services,*[1] and had been hearing more and more about the ecosystem services idea. I decided to attend and invited two colleagues, economist Steven Kraft and geographer Chris Lant, to travel the road to St. Louis with me.

The meeting room was filled with ecologists, economists, other social scientists, and exactly two lawyers—me and a friend of mine from another law school, Jim Salzman. In particular, no attendees represented policy-making institutions such as legislative committees or government agencies. This lopsided turnout troubled Jim and me. Law, after all, eventually has to enter the picture for ecosystem services to be put into operation as a meaningful policy driver. Even more conspicuous was the absence of any representatives from the private sector. For law to put the ecosystem services concept into play, it would be important for landowners, resources users, and participants in business markets to understand what it is about.

Almost ten years later, another major conference on ecosystem services was organized, this time by Jan Davis of the Texas Forest Department and Brad Raffle, a private-sector lawyer. The conference was not convened in the lush greenery of the Missouri Botanical Gardens; it was held in Houston, the heart of the nation's energy and chemical industry. Jim Salzman and I were asked to deliver plenary talks to provide background on the ecosystem services concept, and Geoff Heal was invited to give a keynote talk at lunch. We did not know what to expect. Would anyone be interested in what we had to say? Indeed, would anyone even be there?

To our astonishment, the room was filled with over two hundred attendees, almost none of whom were from academic or research institutions. Instead, the conference was teeming with lawyers, consultants, government personnel, and business managers from over twenty states. The discussion focused not on what ecosystem services are, but on how to "operationalize" them through markets and policies. Needless to say, I was amazed at the ground the ecosystem services concept had gained in less than ten years. What would have been an empty room in 1998 was standing room only a decade later.

What changed? There is no single event to answer that question. To be sure, academics from a wide variety of disciplines have been pushing hard on the ecosystem services concept since the Saint Louis conference. The number of journal articles in ecology, economics, and law has almost doubled each year. The turnout at the Houston conference, however, suggests that something more had happened than policy makers and business managers reading the academic literature. Rather, through a series of events large and small, from Hurricane Katrina down to the parched field of a small farm, people have come to realize that nature is green not in just the ecological sense, but also in the economic sense. Nature, it turns out, is a cash cow.

* * * * * *

This chapter provides background on the concept of ecosystem services—what they are, their context, and the next steps needed to establish them in law and policy. The main point is that it is not only economically useful but also socially *essential* that we fully incorporate the value of ecosystem services in private markets and public policies. We trade and regulate manufactured goods, human-supplied services, and natural resource commodities in vast markets and elaborate regulatory regimes. What we have been overlooking, however, is the economic value that ecosystems provide in the form of services such as flood control, soil stability, gas and thermal regulation, and pollination. We have ignored them because they have appeared to be free—after all, we don't have to pay for photosynthesis. But just as water moves from cheap to pricey as it becomes scarcer, the diminishing stock of "natural capital" has made many of the ecosystem services that flow from it more scarce as well. Increasingly, we are finding that it is costly to go without ecosystem services. It is costly either to replace them with technological fixes (e.g., a stormwater retention pond) or to suffer the consequences of forgoing them altogether (e.g., flooding).

The people from government and business who attended the Houston conference are people who get it. They understand that ecosystem services are anything but free in the long run and that managing them is a matter that should be of great interest to government and business. The questions that I believe will define the next twenty years of natural resources policy are how many more of us will get it, and what will we do about it?

**What Are Ecosystem Services?**

The central thesis of the ecosystem services concept is that ecological resources serve as natural capital for producing not only valuable commodities such as timber, minerals, and water, but also valuable services—ecosystem services—such as water filtration, storm surge mitigation, water recharge, soil stability, and pollination.[2] Implicit in most natural resource laws is the understanding that ecosystems provide a wide range of benefits to humans, including "serving" us life-sustaining benefits.[3] But the expression of that benefit has been at best implicit in law because there has been no scientific or economic foundation on which to build explicit policy goals. The science of ecology focuses on the importance of ecosystem processes in natural contexts, whereas the discipline of economics focuses on the pricing of goods and services in markets. Without information from ecologists about the delivery to humans of ecosystem services, however, the market necessarily will underrepresent their values in pricing decisions. Researchers in both fields have thus begun to fill in the very large hole of knowledge surrounding how *ecologically* important ecosystem attributes provide *economically* valuable services to humans.

With that mission in mind, the ecology literature is burgeoning with efforts to identify and assign value to the service component of ecosystems. Daily, for example, defines ecosystem services as "the conditions and processes through which natural ecosystems, and the species that make them up, sustain and fulfill human life."[4] Similarly, in their classic article on ecosystem services and natural capital values, Robert Costanza and colleagues. define ecosystem services as "flows of materials, energy, and information from natural capital stocks which combine with manufactured and human capital services to produce human welfare."[5] The standard typology for providing more detail to these broad definitions has come from the United Nations' Millennium Ecosystem Assessment (MEA). The MEA is a massive international work program designed to meet the needs of decision makers and the public for scientific information about the consequences of ecosystem change for human well-being and about the options for responding to those changes. With that mission, it is no surprise that the project focuses heavily on ecosystem services, which it groups into four categories: provisioning services (e.g., food and water); regulating services (e.g., wetlands filtering pollution); cultural services (e.g., rivers providing recreation opportunities); and supporting services (services necessary for the production of other service types).[6]

Humans would miss the benefits of those four streams of ecosystem services if they were to disappear. Indeed, we often find it cost efficient to "manufacture" ecosystem services by replicating natural structures, as in the case of so-called constructed wetlands that are built to remove nutrients and sediments from polluted water sources such as municipal wastewater and agricultural runoff.[7] Yet ecosystem service values derived directly from nature show up practically nowhere in our economy as it is structured, and even less in the law supporting that structure. For example, wetlands, it turns out, provide protection against the heat-radiation effect. Heat radiates away from the ground on dry winter nights, rapidly lowering soil temperatures and freezing the moist root zone, but the presence of wetlands in an area keeps the air moist and thus reduces the freezing effect. That service helps prevent crop freezes, but one searches in vain for any recognition of that value in financial or policy marketplaces.[8] That ecosystem services have value is indisputable; however, what that value is and how to account for it in our day-to-day economic and legal decisions are far less clear.

## The Context of Ecosystem Services

Moving from theory to implementing practical applications of the concept of ecosystem services requires that law and policy have a firm grasp of the context of natural capital stocks and ecosystem service flows. It is not sufficient in this regard simply to identify an ecosystem and declare it economically valuable on the premise that all ecosystems produce ecosystem services. Not all ecosystem *functions* yield ecosystem *services*, because the essential feature of ecosystem services is that they are valuable to and used by a human population. Hence, to understand the context of ecosystem services, one must appreciate their *ecological*, *geographic*, and *economic* attributes.

### *Ecology*

Because ecosystems are complex, ecosystem services are complex—the "prototypical examples of complex adaptive systems," according to ecologist Simon Levin.[9] John Holland, one of the leading figures in complex systems research, explains why: "Ecosystems are continually in flux and exhibit a wondrous panoply of interactions such as mutualism, parasitism, biological arms races, and mimicry. . . . Matter, energy, and information are shunted around in complex cycles. Once again, the whole is more than the sum of its parts. Even when we have a catalog of the activities of most of the participating species, we are far from understanding the effect of changes in the ecosystem."[10]

Certainly the basic quality of complex systems exists in ecosystem dynamics, in that "we cannot understand ecosystems only by considering their separate components."[11] Costanza sums up the difficulties of studying such ecological systems: "Complex systems are characterized by: (1) strong (usually nonlinear) interactions among the parts; (2) complex feedback loops that make it difficult to distinguish cause from effect; (3) significant time and space lags; discontinuities, thresholds and limits, all resulting in (4) the inability to simply 'add up' or aggregate small-scale behavior to arrive at large-scale results."[12]

Using this model , ecologists routinely account for two properties as the central products of complex system dynamics operating in ecosystems: *resistance*, the ability of an ecosystem to withstand external stress without loss of function; and *resilience*, the ability of an ecosystem to recover from disturbance.[13] The more that is learned about these properties, the more researchers appreciate that "with complexity comes uncertainty. . . . We must recognize that there will always be limits to the precision of our predictions set by the complex nature of ecosystem interactions."[14] Add to this the nature of political and social institutions involved in ecosystem management as exceedingly complex systems themselves, and the problem of devising ecosystem services policies becomes all the more daunting.[15]

For example, we know that within any given ecosystem the benefits we can confidently identify as services are the result of complex processes associated not only with that ecosystem, but with all linked ecosystems. Therefore, we cannot easily select the services we use for rate, location, combination, or other qualities, as we can, say, the consulting services we purchase from accountants or physicians. Ecosystem services are where they are, and what they are, unless we alter the underlying ecosystem processes.

Altering ecosystem processes is a complicated proposition for two reasons. First, the way ecosystem processes produce services may not be fully understood, so we cannot know with confidence what the effects on service yield may be of changing one or more processes.[16] Second, we know that ecosystems are open, dynamic systems linked to other ecosystems, meaning that to alter the services provided by one defined ecosystem we may need to alter the processes of another ecosystem, which may affect process flows in yet another ecosystem.[17]

Taking these principles to the ecosystem services context, Costanza and colleagues. observe that "ecosystem services and functions do not necessarily show a one-to-one correspondence. In some cases a single ecosystem service is the product of two or more ecosystem functions whereas in other cases a single ecosystem function contributes to two or more ecosystem services."[18] There are, in other words, extensive interactions and an aggregation of diverse elements in play within an ecosystem to produce ecosystem services of value to humans.

### *Geography*

Tracing ecosystem services to human populations is complicated not only by the complexity of the ecological processes, functions, and structures behind the services, but also by the geographic distribution of natural capital that is the source of ecosystem services, of the channels of service delivery, and of the service beneficiaries. In each of these respects, ecosystem services might exhibit patterns that range from being relatively discrete and predictable, making them easy to identify and map (e.g., a boreal forest), to being highly variable in nature, which would complicate their geography (e.g., an ephemeral stream). Of course, any one ecosystem may present a complex landscape of different but interrelated ecosystem services that exhibit a mixture of such properties, making it difficult to pinpoint the geographic context of a single identified service. A forest, for example, might provide shade on a hiking path, soil stabilization to a nearby farmer, water quality to a distant city, and carbon sequestration benefits for all of humanity.

To further confound our ability to provide geographic representations of ecosystem service distribution, the patterns exhibited may vary depending on the spatial and temporal scales of ecosystem processes, the most difficult to define being those involving numerous sources from which service values are most relevant on a cumulative basis. What looks like a discrete source of an ecosystem service on a local scale on an immediate basis may be one of thousands of diffuse sources of a different service operating in unison at a regional level over longer periods of observation. A tree provides shade moment to moment in a defined area; many trees combine to provide the water quality-enhancing properties of a forested watershed. A particular natural capital resource might provide one service over short-term periods (e.g., stormwater retention by a wetland during significant rainfall) and another over longer time frames (e.g., groundwater recharge from the same wetland).

Because ecosystem processes operate over a wide range of spatial and temporal scales, therefore, knowing which set of scales is most appropriate for making particular economic and social policy decisions is no straightforward matter. Indeed, the valuation of ecosystem services is as profoundly influenced by their geography as by their ecology. As Konarska et al. suggest, "The distance of the ecosystem to a population center, the fragmented nature of many ecosystems, the purchasing power of people in various parts of the world, and the spatial scale at which the ecosystem extent is measured, all can influence the valuation of ecosystem services."[19] Farmers in a watershed may see riparian habitat as forgone grazing land. Urban dwellers far downstream enjoying the water quality benefits the riparian habitat provides most likely do not know how the benefits are provided, from where, or when. As geographic scale separates natural capital from the places and times when ecosystem services are enjoyed, valuation can become difficult, which is why we turn next to the economics of ecosystem services.

*Economics*

The ecological and geographic complexity of ecosystem services has several practical implications for the economics of those services. It will usually be the case that the land on which a stock of natural capital is located is not owned by the same person who owns the land on which the ecosystem services flowing from that natural capital are enjoyed. Moreover, it will often be the case that the owner of the natural capital has no way of "charging" the ecosystem service beneficiary for use of the service. To the owner of the natural capital, therefore, the ecosystem service benefits are what economists call *positive externalities*, in that the value is external to the owner's experienced stream of costs and benefits associated with ownership of the property. The farmers mentioned above, for example, don't enjoy the benefits of the riparian habitat they choose to keep intact, but they do feel the costs of doing so in the form of forgone grazing lands.

It will often be the case that multiple beneficiaries of an ecosystem service exist, none of whom can exclude the others from enjoying the benefit. The urban dwellers who benefit from the water quality services of the riparian habitat, for example, all share in them. The benefits of ecosystem services, in other words, often behave as what economists call *public goods*, in that users are hard to exclude and one person's use does not diminish the supply available for other users. Even if the owner of natural capital could get someone who enjoys its benefits to pay for those benefits, that would not prevent others from enjoying the benefits as well. When this is the case, beneficiaries are unlikely to willingly pay for benefits; they are more likely strategically to "free ride" on the continued supply.

The combination of these conditions—the owners of natural capital seeing ecosystem services as positive externalities and the beneficiaries of ecosystem services seeing them as free for the taking—means that ecosystem services will be undersupplied, even though we know they are of vital importance. The owners of natural capital have no rational economic reason for continuing to supply them, and the beneficiaries of ecosystem services have no rational economic reason to pay for them. Clearly, it is imperative that a law and policy of ecosystem services form to make the rational decision the one that sustainably manages our stock of natural capital.

**Next Steps**

Significant challenges lie ahead for the formulation of law and policy for ecosystem services. Critical steps toward that direction, therefore, may include the following:

- *Defining property rights.* Property rights in modern American property law do not merely ignore ecosystem services, but actually impose a disincentive to landowners against keeping natural capital intact. Owning undeveloped natural capital is disfavored in our property law, built as it is on the foundation of converting wildlands to agriculture (and today to suburbia).[20] Owning property also has not traditionally entitled the owner to a continued supply of ecosystem services.
- *Priming markets.* Opportunities for market-based uses of the ecosystem

services concept should be explored at every opportunity, whether through subsidies, tax policy, or tradable rights.[21] Agricultural policy, for example, has depended for decades on crop subsidies that could be transformed into ecosystem service delivery payments. Rural residential development scenarios could use tradable rights to promote conservation of natural capital in return for higher-density, clustered development rights (which in turn help ameliorate sprawl). The point is to direct policies at identifying ecosystem service values in order to prime private markets to do so as well.

- *Designing governance institutions and instruments.* Existing political boundaries and structures are poorly aligned with the multiscale nature of ecosystem services. Efforts to realign political institutions to enable them to develop efficient and effective approaches to ecosystem services will be a politically difficult but necessary step.[22] The transition from current status to markets and policies that integrate ecosystem service values is also likely to present an uneven distribution of "winners and losers," suggesting that political institutions will need to tread carefully in rolling out policies.
- *Identifying research needs.* For any of the above to happen, further research on practical applications of natural capital and ecosystem service theories will be necessary. In particular, trade-offs between ecosystem services and the ecological and economic states of ecosystems need to be more clearly defined. Better understanding of the ecological, geographic, and economic attributes of ecosystem services can only lead to better design of property rights, markets, and governance institutions. Research on ecosystem services should expand to meet those policy needs.

## Conclusion

The need to trace ecosystem services to human populations causes some observers to question the wisdom of engaging in this brand of ecology. They worry that it will "commodify" ecosystems, and they fear that efforts to assign economic values will detract from other policy grounds for ecosystem protection, such as the conservation of habitat for wildlife benefits and the protection of natural systems simply for their intrinsic values.[23] There is no reason, however, that valuing ecosystem services should be seen as a threat to ecosystem conservation. Indeed, quite the opposite seems the case—failure to refine our understanding of ecosystem services and to account for those values in regulatory and market settings, and, more important, in the public mind, is unlikely to promote their conservation.

Those advocating greater policy attention to ecosystem services observe that whether we know their precise economic value or not, the fact that society has to choose how to allocate natural resources necessarily requires valuation of ecosystem services in some form or another.[24] As David Pearce put it, "The playing field is not level; rather, it is tilted sharply in favor of economic development. Two things have to be done to correct this situation. First, one has to show that ecosystems have economic value—indeed, that all ecological services are economic services. Second, a way has to be found to 'capture' the nonmarket values of ecosystems and turn them into real benefits for those who practice conservation."[25] To put it bluntly, it can't possibly help the cause of sustainable ecosystems to sit on the sidelines unwilling to engage in research on ecosystem service values.

On the other hand, neither is unbridled enthusiasm about the development of ecosystem services research and policy warranted. As the foundations for the ecology of ecosystem services form, danger lurks in two respects. First, the new focus on identifying and tracing ecosystem service values may lead to oversimplified conceptions of the underlying ecosystem processes and functions that make them possible. At the same time, a strengthened interest in consuming valuable ecosystem services could lead to management of ecosystem processes and functions for their service values rather than for the sustainability of ecosystems generally. The integration of ecosystem service values in law and policy must happen, but precautions must be taken to ensure that both those potential effects are understood and controlled.

From an idea a few ecologists suggested in the 1980s to a fast-emerging organizing principle of ecological economics in the 1990s and then to a viable natural resources management topic in the 2000s, the ecosystem services concept gained ground quickly. No doubt this is because it simply makes good sense both ecologically and economically. As sensible as it is, however, the challenge will be putting the concept into action through practical applications of law and policy. The science of ecosystem services must stay engaged in that process for it to take hold and succeed, and succeed it must if we are to manage natural resources sustainably into the future.

**Notes**

1. G. Daily, ed., *Nature's Services: Societal Dependence on Natural Ecosystems* (Washington, DC: Island Press, 1997). This multiauthor volume remains the essential starting point for coming up to speed on the science of ecosystem services.
2. For some of the more concise explanations of the role of natural capital in producing valuable ecosystem services, see J. Clark, "Economic Development vs. Sustainable Societies: Reflections on the Players in a Crucial Contest," *Annual Review of Ecology and Systematics* 26 (1995): 225–248; R. Costanza and H. Daly, "Natural Capital and Sustainable Development," *Conservation Biology* 6 (1992): 37–46; and G. Daily and S. Dasgupta, "Ecosystem Services, Concept of," in *Encyclopedia of Biodiversity*, ed. S. Levin (San Diego, CA: Academic Press, 2001), vol. 2, pp. 353–362.
3. As Geoff Heal puts it, almost anyone can appreciate that ecosystem services provide "the essential, low-level infrastructure upon which human activities and built systems rest." G. Heal, *Nature and the Marketplace: Capturing the Value of Ecosystem Services* (Washington, DC: Island Press, 2000), p. 2.
4. Daily, *Nature's Services*, p. 3.
5. R. Costanza et al., "The Value of the World's Ecosystem Services and Natural Capital," *Nature* 387 (1997): 253–260.

6. The most accessible of the MEA's reports are *Ecosystems and Human Well-Being: A Framework for Assessment* (Washington, DC: Island Press, 2003), and *Ecosystems and Human Well-Being: Synthesis* (Washington, DC: Island Press, 2005).
7. For technical and assessment discussions of constructed treatment wetlands, see R. Kadlec and R. Knight, *Treatment Wetlands* (Boca Raton, FL: Lewis Publishers, 2004); D. Steer et al., "Life-Cycle Economic Model of Small Treatment Wetlands for Domestic Wastewater Disposal," *Ecological Economics* 44 (2003): 359–369; and United States Environmental Protection Agency, *Constructed Treatment Wetlands* (2004).
8. These benefits are displayed in a modeling exercise showing the freeze effects with and without wetlands for Florida orange groves in C. Marshall et al., "Crop Freezes and Land Use Change in Florida," *Nature* 426 (2003): 29–30.
9. S. Levin, "Ecosystems and the Biosphere as Complex Adaptive Systems," *Ecosystems* 1 (1998): 431–436.
10. J. Holland, *Hidden Order: How Adaptation Builds Complexity* (Reading, MA: Addison-Wesley, 1995), p. 3.
11. R. Bailey, *Ecosystem Geography* (New York: Springer-Verlag, 1996), p. 16.
12. R. Costanza, "Ecological Economics: Reintegrating the Study of Humans and Nature," *Ecological Applications* 6 (1996): 978–990.
13. Some of the leading work on these concepts can be found in C. Folke et al., "Biological Diversity, Ecosystems, and the Human Scale," *Ecological Applications* 6 (1996): 1018–1024; C. Holling, "Surprise for Science, Resilience for Ecosystems, and Incentives for People," *Ecological Applications* 6 (1996): 733–735; C. Holling and L. Gunderson, "Resilience and Adaptive Cycles," in *Panarchy: Understanding Transformations in Human and Natural Systems*, ed. L. Gunderson and C. Holling (Washington, DC: Island Press, 2002), pp. 25–62; D. Tilman, "The Ecological Consequences of Changes in Biodiversity: A Search for General Principles," *Ecology* 80 (1999): 1455–1474; and R. Virginia and D. Wall, "Ecosystem Function, Principles of," in *Encyclopedia of Biodiversity*, ed. Simon Levin, vol. 2, pp. 345–352.
14. N. Christensen et al., "The Report of the Ecological Society of America Committee on the Scientific Basis for Ecosystem Management," *Ecological Applications* 6 (1996): 665–691.
15. For some of the leading thinking on social-ecological systems, consult M. Janssen, ed., *Complexity and Ecosystem Management* (Cheltenham, UK: Edward Elgar

Publishers, 2002), and B. Walker, L. Gunderson, A. Kinzig, C. Folke, S. Carpenter, and L. Schultz, "A Handful of Heuristics and Some Propositions for Understanding Resilience in Social-Ecological Systems," *Ecology and Society* 11 (2006): 13 (www.ecologyandsociety.org/vol11/iss1/art13).

16. An excellent summary of this property of complex adaptive systems is found in Holland, *Hidden Order*.
17. Bailey, *Ecosystem Geography*, p. 254.
18. Costanza et al., "The Value of the World's Ecosystem Services," p. 254.
19. K. Konarska et al., "Evaluating Scale Dependence of Ecosystem Service Valuation: A Comparison of NOAA-AVHRR and Landsat TM Datasets," *Ecological Economics* 41 (2002): 492.
20. J. Sprankling, "The Antiwilderness Bias in American Property Law," *University of Chicago Law Review* 63 (1996): 519–590.
21. J. Salzman, "Creating Markets for Ecosystem Services: Notes from the Field," *New York University Law Review* 80 (2005): 870–961.
22. J. Ruhl et al., "Proposal for a Model State Watershed Management Act," *Environmental Law* 33 (2003): 929–947.
23. For expressions of this concern, see J. Chen, "Webs of Life: Biodiversity Conservation as a Species of Information Policy," *Iowa Law Review* 89 (2004): 495–608; and H. Doremus, "The Rhetoric and Reality of Nature Protection: Toward a New Discourse," *Washington and Lee Law Review* 57 (2000): 11–73.
24. A strong case for this view is made in Costanza, "Ecological Economics."
25. D. Pearce, "Auditing the Earth: The Value of the World's Ecosystem Services and Natural Capital," *Environment* 40 (1998): 23–28.

CASE STUDY FIVE

# CONSERVATION AT THE SPEED OF BUSINESS

*William J. Ginn*

In July 2005, I was camping with my family on a remote island nestled in Maine's rugged Penobscot Bay. Even in such areas as that, the ubiquitous cell towers now provide at least intermittent service. Although my colleagues at The Nature Conservancy (TNC) are usually pretty good at honoring vacations, the persistent buzzing and beeping of my phone one day finally overcame my reluctance, and I dialed into my voice mail. The news was stunning. International Paper (IP), the largest private landowner in the United States, had just announced that it was going to sell all six million acres of its vast holdings in eighteen states in a single auction in the fall. The persistent calls on my cell phone were messages from my colleagues in states across that ownership asking for guidance on how TNC might use the opportunity to protect IP's treasure trove of high conservation-value forests.

For more than a decade, TNC had worked in partnership with IP to inventory and assess conservation opportunities on the company's land. Who knew what the objectives of new owners would be and whether they would continue the stewardship collaboration developed between TNC and IP over the years?

This was not, of course, the first property sold by IP. In fact, this announcement was simply the final coda in a long, slow process of selling

U.S. timberlands as the company struggled to reposition its business around newer, more strategic opportunities elsewhere. But this announcement was different. Instead of the multiyear process of negotiation that conservation groups had become accustomed to, IP had made it clear that all of its timberlands would be sold in a matter of months. How could the conservation community even be relevant in what would surely be billions of dollars of assets changing hands?

Although the scale of the sale was shocking, we should not have been very surprised. The trend had been clear for years. Integrated forest product companies owning both mills and woodlands had been out of style on Wall Street for a decade or more, and analysts and investors had increasingly been calling for the sale of land as a way to realize greater shareholder value. Behind those calls were two important drivers. First, the U.S. forest products industry had been in a long slump. Demand for paper in the United States was flat, and the forecast was grim—little would change to increase either prices or volume. In the United States at least, stable population growth, recycling, e-mail, and the Internet were creating the perfect storm in the publishing and printing business. What is more, suppliers from Indonesia and Brazil, with cheap land and labor and new, ultramodern mills, were more than competitive. But there was even more pressure from other fronts. The financial good fortune of the baby boomers was increasing demand for rural recreational land, and in many places, it was simply more attractive to grow house lots than timber. All of those trends—development valuations, globalization, and tax efficient buyers—have in less than two decades transformed one hundred years of traditional ownership of forestlands.

As I was the leader of TNC's Global Forest Initiative, it fell to me to help a dozen TNC state programs make sense of the opportunity presented by IP. The task seemed impossible, given the short timeline announced by the company; yet it was impossible *not* to work toward a conservation solution, given the enormous stakes. The good news for TNC was that all of our joint planning and conservation activities with the company over the years meant that we knew a lot about IP's lands. Over a two-week period teams of local TNC scientists worked with regional staff to cross-reference and compile potential areas of interest. In the end the list grew to nearly one million acres of potential conservation interest.

Then it was time to talk to IP. Here again, our long history of investing in that relationship allowed us to pick up the phone and put our audacious request directly to senior managers at IP: Would the company agree to "carve out" from its auction a set of high-conservation properties and give TNC a chance to buy? Initially, the company was reluctant. While they treated our interest with respect, the future of the company depended on delivering on what they had promised Wall Street. For TNC to be a player, it would have to work at the speed of business, not at the leisurely pace of government. Besides, TNC was talking about a deal an order of magnitude greater than any previous conservation purchase. Did we have the money to execute such a transaction?

What followed was a series of intense meetings at IP's new Memphis headquarters to hammer out the potential outlines of transaction. Could we reach a definitive set of business terms in 90 days and close in 120 days? Where was our money to come from? Some initial land proposals were rejected outright by the company as impractical. To be viable we would need to purchase whole tracts, not just "ecological" zones. Other tracts simply were too valuable as part of a larger sale package to be carved out by conservation. In at least one case, the company simply determined that an auction would likely deliver greater value. But the company also brought out detailed ecological maps of its own and urged us to acquire several sites IP thought were simply too important for conservation to ignore.

We needed to be practical ourselves. It was clear that our million-acre package would cost at least a billion dollars, and that was more dollars than we could ever dream of mustering. What's more, conservation groups need to partner with state and federal agencies. In states like Mississippi and

Louisiana, staggering still under the post-Katrina hurricane damage, pulling together the financial resources needed was simply impossible. TNC lost contact with 90 percent of its members in Louisiana when New Orleans was evacuated and federal agencies were focused on disaster aid and not conservation.

In the end, we determined to focus on 240,000 acres in ten states, which in twenty-six different sites represented the best of the best conservation land owned by IP. In addition, we decided to bid competitively on IP's Wisconsin lands in hopes that we would top all bidders on those 62,000 additional acres. Our approach found some champions within IP's forester ranks, who saw our purchase as a way to honor one hundred years of IP forestland stewardship. But an agreement to negotiate was only the beginning of the work. Appraisals were needed; forest inventories needed to be evaluated. And, of course, support would be needed from other nonprofits and state and federal agencies.

A sometimes harrowing negotiation was conducted over a ninety-day period and resulted in a successful agreement. What is more, our bid for Wisconsin's Goodman tracts topped all comers, and suddenly we found ourselves slated to purchase a record $384 million of high conservation-land forests—the largest single private conservation transaction in the history of the country and probably the world. Over the next six months formal purchase and sale agreements were executed, forest inventory cruises were compiled, board approvals finalized and titles checked. I'm happy to report that we did in fact "operate at the speed of business"—in fact, faster than business, as most of our sales were completed before any of the "commercial" transactions.

But how is it that TNC and its partners were able to spend $384 million, especially in light of current levels of federal funding? In the budget year of 2006, less than $40 million was appropriated by Congress for the main land protection program—the Land and Water Conservation Fund. Clearly, we would have to look elsewhere for most of our funding. In the end there were two key parts to our funding strategy: (1) the growth of state and local funding for conservation and (2) partnerships with private-equity partners.

The growth of state funding has been a bright spot in conservation, especially in the South. Led by Florida's long-standing $3 billion land fund, the states south of the Mason-Dixon Line have been passing bond issues and special funding approaches to conserve land. The rise of local and state initiatives is a relatively new phenomenon, but in 2006, over $5 billion of conservation funding was approved by voters. At the core of those investments was a deep concern that the traditional natural resource industries of the states in the region would be undermined by the lost of forestlands. In North Carolina, the governor has established an ambitious goal of conserving 1,000,000 acres of land—in part because the state has lost over 1,000,000 acres of forests to development in the renaissance of Charlotte and other new economic centers. Virginia's governor has a similar, 400,000-acre, goal for his state. In Wisconsin, Governor Jim Doyle showed the way by providing nearly $35 million of state bond money for the purchase of IP lands in that state.

In the end local governments are slated to provide over $200 million of funding—ten times the expected federal funding.

Partnerships with private investors provided $110 million to supplement TNC's conservation funding (figure C5.1). For some, the idea of partnering with private investors may be unsettling, but in truth it represents the largest potential source of new capital for conservation projects. Over the years, TNC has been experimenting with such partnerships but at a far smaller scale. Early on in our work to protect forestlands it became clear that in many cases our conservation objectives could be met by protecting sustainably managed working forests from development and fragmentation. Why not capture the potential value of this stream of income through partnerships with private forestland investors?

Most of the IP project sites could tolerate limited harvesting, and in some cases actively restoring and integrating management strategies such as fire

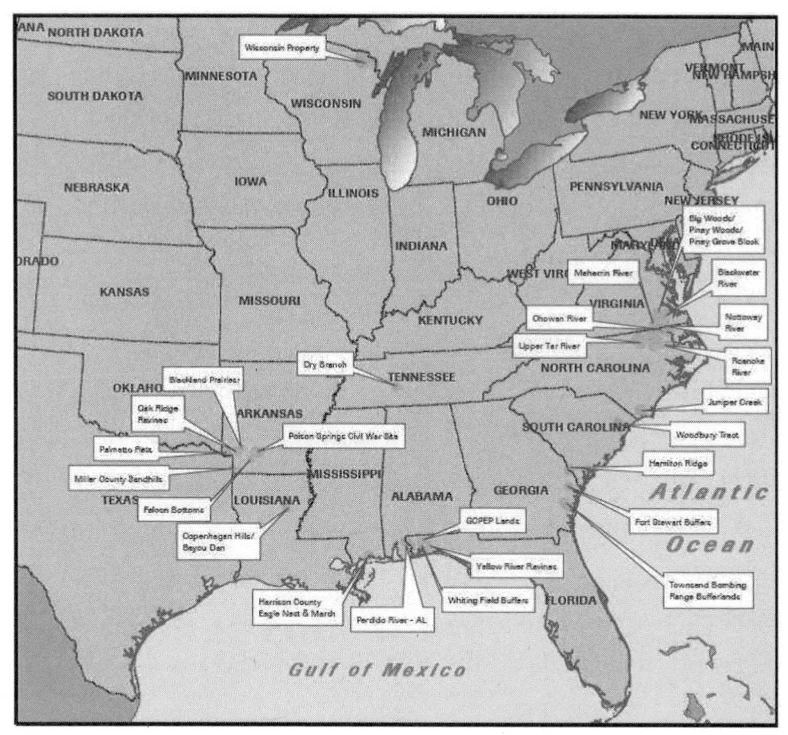

Figure C5.1 The Nature Conservancy–International Paper forestland purchases by 2006.

would be critical to achieving our goals. In addition, some IP projects posed special problems. In one area in Georgia, our plan is to restore longleaf pine, the native species, but currently portions of the site were in plantations. By selling a "timber deed," the right to cut trees for a set period, investors captured the value of those plantations, leaving us the newly harvested land ready to restore. Finally, in several projects, private investors purchased land from us with agreements that would allow us to buy it back in a few years if we could obtain the funding needed over time. Putting in place these complicated relationships in such a short time would not have been possible if we had not been making efforts already to cultivate investors for precisely the opportunity presented by IP. In early 2004, TNC signed a memorandum of understanding with Conservation Forestry LLC, a group led by entrepreneurs

John Tomlin and Paul Young. Under the agreement the duo would provide capital from private investors to conservation-oriented deals developed by TNC. We were able to quickly reach agreements on parts of the IP project needing private capital. At closing, Conservation Forestry, in partnership with another timber investor, Forest Investment Associates, provided $110 million of capital needed for projects in three states.

Another aspect of winning public support for these conservation projects was planning for the potential impact on the region's pulp and lumber mills. It is clear that for every conservation-oriented dollar invested by the states, at least another dollar was motivated by protecting jobs. In part, this was addressed by TNC agreeing to supply fiber where possible to IP's mills. Conservation groups are not accustomed to working as suppliers of wood products but are increasingly being challenged to learn new approaches as we seek to support communities crucial to our success. In the end the case can be made that the best conservation is an economic reason to own and manage forests.

Despite the historic size and scale of the IP project, it is clear that we face a lot more challenges. The other 5.5 million acres were sold conventionally to investors with no conservation protections. A recent U.S. Forest Service report suggests that we will see over 44 million acres of private forestlands lost to development over the next twenty years. If true, that means that we would need to accomplish 146 projects just like the IP project during that time period. A big challenge, but such a goal may prove attainable with the growth of state funding and the new tool of partnerships with private investors.

Including the IP agreements, TNC completed over one million acres of forest conservation projects in the ensuing twelve months. Altogether those projects used an astonishing $800 million of capital, and the private sector bought almost $300 million of that need. Even more tools will be required, but at last we can envision the outlines of what it will take to be successful in saving these critical forest landscapes.

CASE STUDY SIX

# CALIFORNIA NORTH COAST FOREST CONSERVATION INITIATIVE

*Chris Kelly*

Standing with his staff and board members, Sam Schuchat, executive officer of the California Coastal Conservancy, held his breath as logger Rick Hautula thwacked a wedge into the back cut he had just made in the trunk of a towering redwood tree. With a stiff wind blowing ahead of an early fall storm, it appeared that the tree might teeter backward, in the direction of the gathered throng. "THWACK!" Then slowly, almost imperceptibly at first, the big tree leaned away from the crowd of conservationists. Finally gaining speed, it slammed onto the nearby logging road, just as Rick had planned. The group's collective sigh gave way to applause as everyone gathered round the fallen tree and the faller.

How could it be that the staff and board of a state agency whose mission is to "preserve, protect and restore the resources of the California Coast" were celebrating the felling of a redwood tree, the icon of California's North Coast? The answer is evidence of a dramatic change in the theory and practice of land conservation over the last twenty years. That change has been sparked by the marked loss of important natural lands to development and our embrace of the principles of contemporary conservation biology, which hold that enduring conservation requires protecting entire ecosystems and the processes that sustain them.

This dramatic shift in scale and emphasis requires us to go to places we haven't been before—into large, working landscapes where we must consider not only ecological processes and the needs of natural communities, but also the social, political, and economic complexities of human communities. This new terrain requires that we learn new things that are foreign to our traditional focus and that we try new things not usually associated with conservation as we have known it.

The Redwood Region of California's North Coast is one of the richest, and rarest, ecosystems in the world. It is home to keystone species such as the northern spotted owl, marbled murrelet, mountain lion, coho salmon, and steelhead trout. It is also home to some of the most productive forestlands in the world.

Industrial forestry is the predominant land use on the North Coast. In Humboldt County, two timber companies own more than 580,000 acres. In Mendocino County, four timber companies own almost 500,000 acres. These vast holdings were assembled over the last century as the predecessors of the current owners sought economies of scale by acquiring smaller parcels from homesteaders and others who were trading life on the land for the opportunities of California's burgeoning cities.

Recently, the combined pressures of investor expectations, inventory depletion, foreign competition, and regulation have been forcing timber companies to sell or convert their lands to more profitable uses. Consequently, many forestland owners are looking at converting their properties to "higher and better uses" that may yield a greater financial return, such as the rural residential subdivisions and vineyard conversions that are increasingly common on the North Coast.

How can these landscapes be conserved? When confronted with land-use changes of this magnitude, the conventional approach of acquisition and management by a public agency, illustrated in the purchase of the Commander Tract described below, isn't sufficient. We needed a new approach that built upon the extraordinary productivity and biological richness of the Redwood Region and the increasing recognition that maintaining large, sustainably managed forestlands is essential to protecting and restoring forest ecosystems. These factors formed the strategic foundation for our forays into nonprofit forestry in the Garcia River, Big River, and Salmon Creek forests.

The 23,000-acre Commander Tract sat like the proverbial hole in the Mendocino National Forest doughnut. The property was held by a consortium of banks, whose loans to a prior owner had gone bad. Seeking to liquidate the property, the consortium opened it to bidding, generating the interest of a group that sought to develop rural ranches and a hunt club. As the largest contiguous private inholding in the forest, the Commander Tract was critical to the U.S. Forest Service's desire to improve water quality and fish habitat in the Black Butte River and Grindstone Creek watersheds.

On this largely traditional deal, the stars aligned. The State of California had money through a multibillion-dollar parks and open-space bond measure passed by California voters in 2002. The Commander Tract was wholly within the forest boundaries and covered by an existing forest plan. Jim Fenwood, the forest supervisor, was eager for the Forest Service to own and manage this land. The Colusa County Board of Supervisors, which often looks askance at public acquisition of private lands, saw an opportunity to increase hunting and fishing opportunities for its constituents. Looking to the greater good, the State of California made a grant to the Conservation Fund to acquire the property, with the deed going directly to the Forest Service.

In short, the conventional model worked. The Commander Tract is now protected for the people of California, and the Forest Service has assumed the long-term management obligations. But can this model be replicated at the scale necessary to address similar threats on the hundreds of thousands of private forest acres currently at risk? Notwithstanding our success with the Commander Tract, rarely is a public agency in California able to intervene on such large holdings.

So when I stood a year later on a ridge overlooking another 23,000-acre tract of private forest—a property that comprised a third of the Garcia River watershed (figure C6.1)—it was apparent that a conventional strategy like that used on the Commander Tract was not an option. Our team at the Conservation Fund saw that we needed a new approach, one that took into account the unrelenting annual costs of maintenance and management, the need for restoration of habitat for coho salmon and steelhead trout, and the importance of the forest to the future of the local timber economy. The Garcia River property, for sale by the same consortium of banks that sold us the Commander Tract, presented an opportunity to try such an approach that, if successful, might be replicable across the region and enable the protection of North Coast forestland at a scale previously beyond our reach, even beyond our imagination.

The Garcia River spills into the Pacific Ocean one hundred miles north of San Francisco at Point Arena. Like most large timbered properties in the region, the Garcia River Forest was owned by a succession of timber com-

Figure C6.1 The Garcia River watershed.

panies, each attempting to work the forest hard enough to meet the demands of its lenders and investors. This history of intensive industrial timber management left a legacy of depleted inventories of merchantable timber, a network of fragile roads on steep slopes of eroding soils, and miles of spawning habitat for salmon and steelhead clogged with sediments.

However, the redwood forest type that dominates the Garcia River Forest is remarkably resilient and productive: Redwood trees sprout from stumps and produce lumber that is uniquely beautiful, durable, and valuable; and there are few pests or diseases. Further, the Garcia River is recognized by the California Department of Fish and Game as a high priority for protection and recovery of the state-listed coho salmon and steelhead trout.

Reflecting on the unique resource values of the property, it occurred to us that while there was presently little timber income, there might just be enough to sustain a not-for-profit owner who could reduce expenses and raise public and philanthropic funds to support the management necessary to rebuild commercial timber inventories, upgrade roads, and restore stream conditions. In other words: Could a large, understocked tract of coastal forest be returned to sustainable timber production and ecological vitality through patient management by a nonprofit organization in partnership with private and public agencies and community stakeholders?

To test this idea, the Conservation Fund partnered with the California Coastal Conservancy, the Wildlife Conservation Board, and The Nature Conservancy to acquire the 23,000-acre Garcia River Forest in 2004 for $18 million. That purchase established the largest nonprofit-owned working forest in the western United States and represented an unprecedented evolution in the role of the participating state agencies. Our goal is to demonstrate that the nonprofit ownership and management of a large, industrial timber property is a viable conservation strategy in a region where current timber values and land-use trends instead encourage subdivision.

Given the unprecedented nature of the project and the controversies surrounding forestry in the region, our funders appropriately asked that we prepare a forest management plan describing our goals and the specific actions to achieve them. Working with the local community, public agencies, and nonprofit partners, we completed an Integrated Resource Management Plan for the forest. The plan establishes the following broad goals:

- Improve ecological conditions for selected conservation targets.
- Increase commercial timber inventory by 300 percent by 2055.
- Generate sufficient revenue from timber management and other sources to cover all costs of management.
- Improve land health through adaptive management.
- Support the local community by hiring local contractors and suppliers.
- Provide public access, educational, and recreational opportunities for the local community.

In November 2006 we began our first timber harvest operation, with broad support from the local community. Since our purchase of the forest in 2004, we have contributed more than $750,000 to the regional economy.

Having established a precedent for nonprofit ownership and management of working forests, we turned our attention to finding sources of funding to replicate this strategy elsewhere in the region. The nexus between forest management, water quality, and coho salmon protection and restoration led us to consider funding sources for water quality projects and programs, one of which is the State Revolving Fund (SRF).

The SRF is a low-interest loan program established under the federal Clean Water Act (CWA) and administered by the California State Water Board to fund a wide range of water quality projects. The obvious appeal of the SRF is that it provides fully amortized loans priced at 50 percent of the state's general obligation bond rate for periods of up to twenty years. Historically, the SRF has been used to fund publicly owned facilities for wastewater treatment. Our review of the program guidelines revealed, however, that the SRF also could be used for projects that address nonpoint source pollution, such as sedimentation related to current and past forest management.

In 2005, we approached the state water board and suggested that a low-interest SRF loan to the Conservation Fund for forest acquisition and management of the type planned for the Garcia River Forest would be an efficient way to achieve some of the board's water quality objectives for the North Coast. We demonstrated that such a loan—in a funding package that included traditional public and private grants—would allow us to protect and enhance water quality by reducing harvest levels, building forest inventory, and converting to uneven-aged forestry. In June 2005, the state water board agreed and awarded eligibility to the Conservation Fund to borrow up to $60 million to acquire working forestlands and improve water quality through changes in management and silvicultural practices.

In 2006, the Conservation Fund entered into an agreement to purchase 16,300 acres of forestland located in the Big River and Salmon Creek watersheds from Hawthorne Timber Company for $48,500,000. To determine the appropriate amount to borrow from the SRF for the purchase, we first worked with our prospective funders to develop forest management guidelines that we believed would achieve our habitat protection and water quality enhancement goals. We arrived at the following general guidelines:

- Apply uneven-aged management using primarily single tree selection.
- Harvest less than 35 percent of standing conifer volume per entry.
- Limit harvest reentry to ten years or greater.
- Limit harvest in riparian areas.

We then applied the agreed-upon management guidelines to the properties' timber inventories and age-class distributions to determine the permissible harvest levels and resulting annual harvest volumes. Finally, we calculated the net revenues realized from the permissible harvest levels. We determined that we could harvest approximately 4,500,000 board feet per year (a 45 percent reduction from the levels allowed under California's Forest Practice Rules), resulting in a net cash flow of approximately $2 million.

We concluded from this analysis that we could meet our project goals and repay a $25 million loan from the SRF. With additional grants from the California Coastal Conservancy, the Wildlife Conservation Board, and others totaling $23,500,000, we completed the purchase on November 1, 2006. As in the case of the Garcia River Forest, the Conservation Fund's goal is to develop a comprehensive management plan for the properties. In the interim, general management guidelines are defined in a memorandum of understanding.

It will take decades to steward these properties from their current conditions reflecting a legacy of clear-cutting and high-grading to a condition of economic stability and ecological integrity. Along the way we will confront many challenges, including relatively low current timber volumes; the unnatural predominance of hardwoods in some stands; the burden of maintaining and improving an extensive road system; and the timber economy's uncertain economic, regulatory, and political environment.

But success is possible. There is broad awareness that North Coast forests sit at a historic crossroads. One road leads to fragmentation, loss of forest productivity, and diminished ecological integrity. The other leads to intact watersheds, recovering fish and wildlife, and a sustainable timber economy for the region. That is the road we choose. It is our hope that the community, with public and private stakeholders, will walk it with us.

CASE STUDY SEVEN

# RANCHING FOR FAMILY AND PROFIT

*Kay and David James*

People, land, and economics cannot be separated. For success in any arena of life, one must see that he or she is part of a whole. A business or a nation cannot exist without a consciousness of the people it touches, nor can it ignore economics in the largest sense—business economics or the healthy economics of the land it occupies. This applies to a one-room business as much as it does to a national identity. History is full of examples of individuals and entities that in their quest for land, wealth, power, and economic dominance have ignored the people they touch or the space they occupy. When human dignity or worth is diminished, the land and the economy will eventually fall into misuse, as well. You may say, what has this to do with "ranching that pays"? A farm or ranch will be as successful as its awareness of the "whole" it touches—its people; its soil, water, and air; its animals; its customers; its neighbors.

It is generally agreed that it takes capital to get into ranching or farming—that all businesses and organizations are dependent on capital. It is seldom recognized, however, that all capital investments are not financial. Since 1993 the families at James Ranch in Durango, Colorado, have been forging a model in which people are our top priority and our most important form of capital investment (figure C7.1). For example, our Navajo cowboy who rode the summer range for forty-five years retired recently, at eighty-one years of age. We hope to retain half of what he knows about the land and animals. People determine the condition of the land and its economics. However, people do not exist or thrive separately from the land or economics—hence the concept of the whole.

Figure C7.1 The James family of Durango, Colorado.

At James Ranch five sources of capital make up our whole:

- *Human capital.* The people of the ranch community are the human capital. Their talents, work ethic, creativity, self-governance, values, and happiness determine the success of every level of activity.
- *Intellectual capital.* Education and knowledge, gained through the life experiences of each member, bring to the ranch the wisdom and organization it needs to be successful.

- *Financial capital.* The land, money, buildings, livestock, equipment, grass, forests, wetlands, land development potential, land easement potential, and estate planning make up the financial capital of the ranch.
- *Environmental capital.* Healthy soil, clean air, clean water, abundant wildlife, and open space enrich the ranch, its families, its employees, and the community.
- *Social capital.* Social capital consists primarily of happy people. All the individuals touched by the ranch and its operations gain physical and emotional health from good relations, local governance, a secure and healthy food supply, open space, and respect for the common good.

You have just read the *why* and *what* of life at the James Ranch. Now *how* do we execute our vision for the whole on a daily basis and make it pay?

Human capital is the most important. Of our five children, three are presently living at the ranch because all our children were made welcome and could see opportunity and encouragement there for creativity and entrepreneurship. Today, our grandchildren are growing organic veggies, flowers, and meat products to sell in their own farm market booth in Durango. We strive to see the uniqueness of each individual and to support the education and development of what makes them happy—and, in turn, a contributing member of the whole, regardless of age or talent (remember our Navajo cowboy). From our home base of 450 irrigated acres, we direct-market landscape trees, pastured eggs, organically grown vegetables and fruits, aged artisan cheese made from Jersey cow raw milk, natural pork, and grass-finished beef. Presently, four families are marketing products to the local community from our on-site farm stand, two farmers' markets, and natural food stores and restaurants.

We encourage education, travel, and activities that contribute to the growth and happiness of the individual and inevitably bring new and valuable ideas to the ranch. We have established a family governance system, which includes regular quarterly family business meetings that harmonize and expedite the sharing of ideas and their implementation. The two families not presently living and working on the ranch attend the meetings and contribute their expertise to the management of the ranch. The running of the meetings is rotated each quarter, so all have an equal opportunity to participate and learn. The grandchildren attend the meetings and contribute in areas important to them. Wisdom, our intellectual capital, is mined from these meetings for the good of the whole. The financial assets and short- and long-term planning for James Ranch are managed by the family and enhance each business and the land itself.

How did we get to this point? After seventeen years in the cattle business (commercial cow and calf, purebred, marketing, hybridizing) we split off a hundred-acre corner of the ranch and developed and built a residential community. Our personal and environmental values created the first planned unit development in our county, which is still held up as a smart example of residential development. That project took twenty-eight years to complete but allowed us to own the remainder of the ranch debt free.

Near the end of that adventure we were yearning to be back in agriculture. Gratefully, we were introduced to four life-changing forces: holistic management, grass-based ranching, direct marketing, and our children returning to the ranch with their families. Now those families independently own and manage their ranch-based businesses, taking advantage of the exciting movement back to natural and organically raised foods and plants. All of our products are raised without petroleum-based fertilizers, pesticides, or herbicides; and the food sold to our customers is free of antibiotics and growth or milk enhancers. We raise our animals and plants as close to nature as we can, and we direct-market our "story" (the story of what we do and why) and our products to our local and regional communities. We are practitioners of relationship marketing; we encourage customers to experience the ranch, to see how their food is produced and meet us—their producer and neighbor.

We have a beef customer who came to our Telluride Farm Market booth four years ago, bought beef all summer, and the next year ordered a half beef—driving from his home in Phoenix, Arizona, to pick it up and tour the ranch. This last summer he ordered another half and brought a friend with him when he picked up his beef. They returned for another half in the fall. This is marketing at its best.

The families at James Ranch practice *minimum input agriculture.* We want to raise animals that make us a living—not to make a living for the animals. For example, female bovines that fail to wean a calf each year are sold. This is a hard thing to do when you really like the looks of the cow and are tempted to forgive the loss of her calf. But the fact is, you have a "zero" for the cost of carrying her for the year and no profit from a calf. Animals are selected for their adaptability to our environment and management practices. In the same vein, we want the minimum amount of equipment and vehicles necessary to function properly. We choose to lease U.S. Bureau of Land Management and Forest Service lands to run our cow and calf operation, rather than servicing debt to purchase land.

Rest rotation and intensive grazing are critical to our management of the irrigated pastures of the home ranch and on our leased lands. The use of portable electric fence on the irrigated pastures allows us to put a grass finish on our beef cattle that rivals feedlot gains and at the same time allows the animals to do the fertilizing, in addition to building humus in the soil by root-shedding each time the grass is clipped with grazing. The results on our government-leased land are less dramatic but no less important. Employing rest and rotation and herding techniques on the summer pastures, and dormant season–grazing on the more brittle lands, gets away from continuous grazing and the "turn-'em-out-round-'em-up" ways of a hundred years ago. Slowly but surely, health will be restored to these lands as we strive to emulate nature in our use of them.

Our ranch in the north Animas Valley is one of the last working ranches in the valley. Because our families call this place home, they value all the assets it represents.

The love and respect that promote healthy people and family relations on the ranch ripple out into the community. Healthy families grow healthy communities, and they, in turn, take good care of the environment. The direct marketing of locally grown products opens lines of communication and respect. Money stays in the community, and people know their vendors and their vendors' values. This is building personal, community, and food security from the bottom up.

At James Ranch "ranching that pays" is about more than paying the bills and the banker. A big part of our pay is seeing the soil full of microbial life; the grass healthier every year; the slick, fat cattle more numerous; the irrigation water filling with living organisms from flowing through our land; the raptors increasing; our grandchildren swimming in the ponds; customers singing the praises of the food we raise; and the peace and awe of a full moon rising after a full day's work.

Return on our investment? You bet!

# PART III

## *The Radical Center: Finding Common Ground*

Part III examines how the ideas of relationships, collaboration, and cooperation prevail in today's natural resources management. Perhaps more than any other topic, cooperation infuses today's conservation efforts. Largely in response to decades of fighting and litigation, contemporary conservation efforts reflect changes in how society approaches conservation challenges. People are increasingly meeting in the middle and abandoning the traditional entrenched sides, both right and left, from which they historically launched attacks. Richard L. Knight's opening chapter in this section argues that food, wood products, and open space are stitching together the traditional opposing sides of rural and urban communities. The relocalization effort in the United States has arisen in response to many things, including concerns about what we eat and where our food and wood products come from, how our communities grow and develop, a globalizing economy, and an energy dependence on countries that may use energy as a geopolitical tool.

Courtney White, cofounder of the Quivira Coalition in Santa Fe, New Mexico, next addresses the topic of land health and ecological restoration in chapter 10. Stewardship of the lands that sustain us, economically, ecologically, and aesthetically, makes sense and is part of the growing relocalization movement. Understanding that healthy lands provide important ecological services and that helping to restore degraded lands offers people opportunities to reconnect with the fundamentals of soil, plants, animals, and water is changing our approach to watershed-based conservation efforts.

In chapter 11 Peter Forbes, founder of the Center for Whole Communities in Vermont, contributes a discussion of the importance of relationships. Forbes challenges the reader to honestly assess people-land relationships in a rapidly changing world. He believes it is "no longer possible to protect land and nature from people." Conservation that works in the decades to come will be mostly about efforts to rebuild connections between people and land, relating land to everyday human life. As we become an increasingly "landless" people, experiencing ever-increasing distances between us and the fundamentals of nature, the need for such connections becomes more urgent.

Part III ends with five case studies that demonstrate that cooperation and conservation are not only compatible but also synergistic in the energy they generate, for healthy human communities and for restoring degraded ecosystems. The subjects of the case studies include a Kenyan seascape, a farm in Iowa, an Oregon watershed, a Montana valley, and the eye-opening experience of Dan Dagget, a former activist who now believes in the necessity for finding common ground among sides that have historically fought rather than cooperated.

CHAPTER NINE

# FOOD AND OPEN SPACES: BRIDGING THE RURAL-URBAN DIVIDE

*Richard L. Knight*

Conversations about conservation can be depressing. Anyone involved in environmental issues knows this well. You meet a friend on the street, in the grocery store, or for drinks after work. Not many minutes into your chat, one of you brings up a recent story or firsthand experience of loss: loss of wildlands, inappropriate development, contamination of water, unhealthy food, increased soil erosion, rampant materialism, climate change, the industrialization of food production. . . . You know what I mean, depressing.

So it was a few days after Gary Paul Nabhan had visited my wife and me that it occurred to us that although Gary had been a house guest for three days and we had spent considerable time together over meals, hiking, at book events, and swapping stories and laughs with friends, we never once had a depressing conversation. Everything we discussed was hopeful. What's going on here? —Ahhh, I get it!

Gary, a prize-winning essayist and conservation biologist; a MacArthur "genius" and Pew Fellow; a cofounder of Native Seeds/SEARCH; and the author of books with such delightful names as *The Desert Smells Like Rain*, *Cultures of Habitat*, and *Coming Home to Eat*, should have been like the rest of us: pessimistic. After all, the earth is in serious trouble—our home planet's atmosphere is destabilized; dire warnings of pending doom are commonplace; the climate is running amok; our food, water, and air are poisoning us; and species extinction has gone exponential. Gary knows all this—he's as tuned in as any conservation biologist could be. Why wasn't his mood a bit on the gloomy side, like that of the rest of us?

The reason Gary's outlook on conservation was sunny rather than gray was this: He deals day to day with the connections between people, open space, food production, and the relocalization of people and land. And, on those issues, the news is bright. Americans seem to be rediscovering once vibrant connections between food, fiber, and open space. Those connections make economic sense in an age of peak oil, make health sense in an age of obesity, and make ecological sense in an age of ecosystem services.

### The Red-Blue Divide

Although Gary's work is about building bridges between rural and urban America, one might arrive at the conclusion after reading newspapers and watching the evening news that America is made up of two opposing sides, mirror opposites that detest each other, a nation divided. Whether the issue is gun control, abortion, same-sex marriage, endangered species, immigration—whatever the flavor of the month—we have become a country that is more about what we disagree on than what we hold in common.

In understanding the conflicts that divide us, we are compelled to look at the now infamous maps of red and blue America. During national elections, Americans seem to divide into two camps: a red Republican group and a blue Democratic group. Although political pundits have made much of the red and blue states, it is more interesting to look at red and blue counties. Metropolitan counties in America tend to be true-blue Democratic; the rural counties tend to vote red-Republican. This rural-urban divide mirrors the division between ranchers, farmers, and foresters in rural America and our country's majority population, anchored in urban and suburban centers.

This rural-urban divide is more than just clashes over values. Importantly, it also exists over the fundamentals: the food and other natural resources we consume and the land on which they are produced. City people want rural landowners to protect wildlife habitat, open space, and ecosystem services, while rural landowners feel that city people take those societal benefits for granted, bestowing not even so much as a thankful nod on their rural neighbors for growing food and producing wood products. Agriculturists are mystified that urban publics aren't more willing to pay a fair price for the food and fiber they do produce; urban people seem mainly interested in shopping at box stores for food at the lowest possible price, neither knowing nor caring where it was produced, how many miles it traveled, or how many hands have touched it. Aldo Leopold wrote, "There are two spiritual dangers in not owning a farm. One is the danger of supposing that breakfast comes from the grocery, and the other that heat comes from the furnace."[1] He said similar things regarding wood products: "The long and short of the matter is that forest conservation depends in part on intelligent consumption, as well as intelligent production of lumber."[2]

The economic reality is that the efforts of farmers, ranchers, and foresters to produce food and fiber are increasingly placed at risk by our global economy and by cheap food and wood products from offshore. The rift between America's rural and urban societies can be overcome only when we appreciate what each produces and the natural interdependencies that bind us. Regardless of one's politics, buying locally makes a lot of sense. Locally grown food is fresher and tastes better. Because local food spends more time in the ground or on a tree, not on an ocean-going ship or long-haul diesel, it arrives on your plate having had longer to ripen, while also having expended less energy to transport. Eating locally, if we did it more, has the mind-calming capacity to reconnect us with the seasons of the watershed we inhabit. Imagine having tomatoes one-fifth of the year rather than every day of the year—wouldn't they be more welcome then (figure 9.1)? And, of course, when you eat locally, you support your rural neighbors who are struggling to keep the land on the edge of town in open space and out of sprawl.

Figure 9.1 The bounty of the land that sustains us all.

A more careful look at a red-blue map reveals further subtleties. Demographers provide us with a third color when examining the rural-urban, liberal-conservative divide: purple. When you mix red and blue, you get purple. And purple, a blending of conservative and liberal expressions, is becoming the dominant color as more and more people question the agendas of both the far-right and far-left ideologies. Does either side care about the well-being of our nation as much as they seem to care about defeating the other, whether in elections or on the public stage?

Just as suburbs complicate what is meant by "urban," exurbs make it increasingly difficult to neatly categorize "rural." For the first time in over a century there are more Americans leaving metropolitan areas to live in the country than there are rural people moving to the cities.[3] Americans are buying small acreages on what were recently farms, ranches, and private forestlands and living exurban existences on ranchettes. Demographers might say, therefore, that the rural-urban divide is passé; today's fault line might more appropriately be called the suburban-exurban split.

### Stitching the Rural-Urban Divide Back Together

Despite attempts by ideologists to create rural-urban dichotomies, the alleged differences are, at their core, false. All Americans prefer healthy food and sustainably grown wood products, and we prize open space. Farmers, ranchers, and foresters produce both, though they are compensated only for the food and fiber, not the open spaces and ecosystem services that their private lands provide to everyone. That more and more of our food and fiber come from offshore and that we are content to shop at box stores for the highest-quality cut of meat for the lowest price make the job of agriculturists and foresters, who guard our nation's private open spaces, more tenuous.

Likewise, rural producers need to acknowledge the importance of cities. Cities are where their customers are, along with libraries, hospitals, universities and colleges, county courthouses, retail businesses, and the factories that build pickup trucks and machines to work the land. If there are true differences between rural and urban in America, they reside in a bubble that will soon burst. We would have to exist in a state of extreme denial to believe that the fates of rural and urban communities are not inherently entwined.

The answer to this rural-urban disconnect is to bridge local food and fiber production to local open space and wildlife habitat—economically and ecologically. This will happen to the degree that Americans choose to eat locally produced food that is well husbanded on land that is well stewarded. To dismiss this solution as nostalgic misses the point. We know that the Earth's soil, water, atmosphere, plants, and animals cannot sustain food production as presently practiced under the agribusiness model. To accept an alternative model of agrarianism is to quicken the arrival of the day when American communities may once again realize their interrelatedness. What is needed today is locally produced food and other natural resources produced on local open spaces, offered and received with grace and a fair market value by urban people who no longer take for granted the societal services of local farmers, ranchers, and foresters.

Wendell Berry captured this need when he wrote, "The most tragic conflict in the history of conservation is that between environmentalists and the farmers and ranchers. It is tragic because it is unnecessary. There is no irresolvable conflict here, but the conflict that exists can be resolved only on the basis of a common understanding of good practice. Here again we need to study and foster working models: farms and ranches that are knowledgeably striving to bring economic practice into line with ecological reality, and local food economies in which consumers conscientiously support the best land stewardship."[4]

## Finding Common Ground

Good news abounds regarding this illusionary divide between urban and rural people. Urban people line up in droves to approve ballot initiatives that protect open space, thumbing their noses at our elected officials who fail to see the high esteem that all Americans place on open lands. Between 1998 and 2007, voters passed 76 percent of 1,547 conservation ballot measures, which generated $44.5 billion for protecting open space (www.landvote.org).

The number of land trusts holding conservation easements is keeping apace with dollars for open-space protection. In 1981 there were 370 land trusts in the United States; today there are over 1,700, which have protected more than 37 million acres (www.lta.org). On average, a new land trust forms in the United States every two weeks.

Ranchers and farmers are getting in on the act, proving beyond a doubt that they would rather farm and ranch than sell out to a developer. In Colorado, the statewide cattlemen's association has created the Colorado Cattlemen's Agricultural Land Trust (see case study 4). The first land trust in the nation to be formed by a group of mainstream agricultural producers, it now holds easements on over 300,000 acres of ranchlands and farmlands. As demonstration that this sentiment among ranchers and farmers is region wide, seven other state cattlemen's associations now have land trusts for agricultural producers (www.maintaintherange.com).

Ranch families have expressed the ultimate private property right, denying forever anyone's "right" to develop his or her most productive and best watered lands, those with the deepest soils and the lower elevations.[5] These ranch and farm families have drawn a line in the soil that speaks to their commitment to land, family, community, and the production of homegrown food. In voluntarily conceding forever the last cash crop of their farms and ranches, these individuals illustrate the differences between takers and caretakers. They represent what so many agriculturists and foresters have always believed: They have never felt like they *owned* the land; they feel like they *belong* to it. More and more of these producers realize that along with their property rights come obligations and responsibilities—to the land, to their neighbors, to the watershed. Perhaps that's why so many private lands are appearing in conservation easements.

Look anywhere in urban America for evidence that urban people and agriculturists are both doing their share to forge a intelligent consumption-intelligent production equation. For example, the organic revolution has arrived. Organic products are sold in over twenty-five thousand natural food stores and cooperatives and in nearly 73 percent of conventional grocery stores. The U.S. organic food market is growing by more than 20 percent each year, and reached a value of over $31 billion in 2007. Students at my home institution of Colorado State University are harking back to the century-old land-grant message in ensuring that what students eat in our dormitories comes from local farmers. Indeed, the students have their own garden and farmers' market, as well as an "eating locally" group.

Some agriculturists are taking their message to town, as well, demonstrating that producing food for local communities is good for the bottom line. These risk-taking individuals can be found almost everywhere, though they are clearly not the dominant voice of American agriculture—yet. By stepping out of the agribusiness box, they have become symbolic of what is possible, and they are helping to demonstrate that ranching and farming, done right, have the ability to be culturally robust, economically viable, and ecologically sustainable.

The number of farmers' markets is now doubling every ten years. In 1994 the United States had 1,646 markets selling locally produced food; ten years later the number had swelled to 3,700. The updated directory now lists over 4,300 markets (www.localharvest.org). Berry captured the spirit of eating locally and supporting farmers' markets when he said, "Farmers' markets and community-supported agriculture are where producers and consumers are meeting each other, face to face."[6] The increase in community-supported agriculture (CSA) ventures is keeping pace with that of farmers' markets; there were estimated to be only fifty CSAs in 1990; today they number over a thousand.

The United States now has over 170 slow-food chapters coast to coast (www.slowfoodusa.org). Members taste, celebrate, and champion foods and food traditions important to their communities. Montana is tapping into this movement; it ranks first nationwide in organic wheat production and second in production of other organically grown grains, peas, lentils, and flax. Working through companies like Great Harvest and Wheat Montana, Montana farmers are growing healthful food and selling it locally. The same trend is emerging for grass-fed beef. Just ask Kay and David James, Colorado ranchers (see case study 7). By producing grass-finished beef, they provide consumers with a healthful and safe product produced on open lands, both private and public.

An important benefit of the relocalization movement that is sweeping the country is alignment with the concept of becoming part of a place. Wallace Stegner thought long and hard about what it means to become native to a place. He wrote, "The frontiers have been explored and crossed. It is probably time we settled down. It is probably time we looked around us instead of looking ahead. We have no business, any longer, in being impatient with history. . . . Plunging into the future through a landscape that had no history, we did both the country and ourselves some harm along with some good. Neither the country nor the society we built out of it can be healthy until we stop raiding and running, learn to be quiet part of the time, and acquire the sense not of ownership but of belonging."[7]

Among many good things that come with relocalization, another benefit is the support and demonstrated respect for the local cultures that have kept rural working lands healthy and productive. As Berry explained during a question-and-answer session at the annual meeting of the Quivira Coalition. "As important a reason as any to support ranching, farming, irrigating, and logging is that our society will need them as teachers, mentors, and critics in the years to come."[8]

These trends of reconnecting people and land through food and open space are all growing, and they point in the right direction. They capture the necessity that David Orr wrote about in *The Last Refuge*: "For agrarianism to work, it must have urban allies, urban farms, and urban restaurants patronized by people who love good food responsibly and artfully grown. . . . It must have farmers who regard themselves as trustees of the land that is to be passed on in health to future generations. It must have communities that value farms, farmland, and open spaces."[9]

Perhaps each of us should take a rancher, farmer, or forester to lunch and discuss food, fiber, open space, and human and land health.

**The "Radical Center" and Home, Land, and Security**

During the heat of a recent political campaign, Americans may have missed two pronouncements. First, the Department of Agriculture announced that, for the first time, Americans imported more food than we exported. Second, the secretary of health and human services warned that terrorists could easily reach across our oceans and hurt us through our imported food.

That we now buy more food from offshore than we grow onshore is the inevitable consequence of our globalizing economy. The secretary's "code orange" alert was quickly pooh-poohed by the White House—all is well in the American homeland. Or is it? Orr wrote, "The resilience that once characterized a distributed network of millions of small farms serving local and regional markets made it invulnerable to almost any conceivable external threat, to say nothing of the other human and social benefits that come from communities organized around prosperous farms."[10]

Political campaigns of recent years have been full of rhetoric about manufacturing jobs going offshore but nary a word over the loss of agricultural and forestry jobs to other countries. The covenant that once bound rural and urban Americans has been severed. Until recently, urban people knew and supported the sources of their food and fiber. Rural people, in return, provided food, wood products, and open space, as well as places for city people to hunt, fish, and visit.

In this light, our government's concern over "homeland security" misses the most important point. A secure homeland is not based on military might. Home, land, and security come together when urban people realize that ecologically sustainable food and fiber production are possible and that rural cultures matter, and when urban people are prepared to compensate farmers, ranchers, and foresters for food and fiber, as well as for protecting open space, wildlife habitat, and watersheds through husbandry and stewardship. Equally important to this winning equation are rural people who acknowledge the importance of urban areas and offer a friendly handshake to their urban neighbors. And, collectively, when rural and urban people are thankful, daily, for the blessings of plants and animals that provide subsistence to us all, and to the lands and waters that sustain us spiritually and aesthetically.

A few years ago I was fortunate to work with a talented group of Americans in establishing a movement that we called the radical center. The phrase, coined by Arizona rancher Bill McDonald, was our coda. What we were proposing was radical because it focused on cooperation rather than the litigious business of conflict and confrontation. It was centrist in that we shunned the trite values of the far right and far left, whose animosities ensure stalemate from their deeply dug trenches. We were all gratified that our collective hero, the late Wallace Stegner, seemed to endorse our efforts with these words: "I have been convinced for a long time that what is miscalled the middle of the road is actually the most radical and most difficult position—much more difficult and radical than either reaction or rebellion."[11]

In a nutshell, though we were as diverse as ranchers, academics, conservationists, writers, historians, artists, and economists can be, we shared a common attribute of being communitarian in nature rather than contrarian. From different life experiences, we had each come to the same conclusion: It is more effective, and more fun, to work on things that unite us, rather than fight over things that divide us.

So in January of 2003, twenty individuals came together in Albuquerque, New Mexico, for forty-eight hours to figure out a way to express our rejection of decades of divisiveness and acrimony that jeopardizes what we all love and value. But we also met to take our country forward, to restore ecological, social, and political health to a landscape that so desperately needs it. We met to write a declaration and ended up writing an invitation instead, an invitation to join the movement that is widely circulated and whose list of signatories is still growing (www.quiviracoalition.org).

You can imagine our delight following the 2006 general elections when *Time* magazine (November 20, 2006) ran the headline "Why the Center Is the New Place to Be." Not that any of us took credit for the growing change across the country; we were just happy that as a nation we were moving in the direction of cooperation and away from conflict.

The radical center has been helpful in my academic teaching, as well. At Colorado State University I am fortunate to teach the capstone course in the College of Natural Resources. Historically, in the syllabus for this course, "combat biology" was listed as one of the objectives. The radical center has empowered me to replace that tradition with a more hopeful objective of working together to promote healthy natural and human communities. I have noticed the students gravitate toward the new approach, as it offers an alternative to the "defend nature to the last person" mentality that has for too long characterized my field. Rather than being the "bad guy" and telling people what they can't do, our students are looking forward to a future in which they work with people on things they *can* do—together.

To find evidence that this radical centrist position is tenable is to look at almost any watershed today. All across the United States, Americans are coming together to cooperate over improving the economic vitality of watersheds while simultaneously working to steward degraded lands back to health. In the words of the radical center, "We have two choices before us. One is to continue the heated rhetoric of the far right and the far left, spending our time slinging insults and hardening polarization. Or, we can join with the hundreds of watershed and community-based programs around the country and move to the radical center. The point where people will respectfully listen, respectfully disagree, and in the end, find common ground to promote sound communities, viable economies and healthy landscapes. By following this behavior we will, over time, so marginalize the wingnuts on the right and left that our voices will be heard over theirs."[12]

To return to reality in the United States will require what Aldo Leopold warned us about in 1928: "Even the thinking citizen is too apt to assume that his only power as a conservationist lies in his vote. Such an assumption is wrong. At least an equal power lies in his daily thought, speech, and action. . . . But most problems of good citizenship in these days seem to resolve themselves into just that. Good citizenship is the only effective patriotism, and patriotism requires less and less of making the eagle scream, but more and more of making him think."[13]

## Conclusion

There are many things the United States needs right now, but there is nothing it needs more than healthy rural lands supporting healthy rural communities with the full awareness and commitment of those living in healthy cities. That farming, ranching, and forestry will survive is not the question. The question is whether those occupations will rediscover their abilities to once more encourage vibrant connections between urban and rural communities, focusing on the rural lands of America that bind us all to this remarkable country. For security's sake, it is time to beat those red and blue swords into purple plowshares.

### Notes

1. A. Leopold, "Good Oak," in *A Sand County Almanac and Other Essays* (New York: Oxford University Press, 1949), p. 6.
2. A. Leopold, "The Home Builder Conserves," in *The River of the Mother of God and Other Essays by Aldo Leopold*, ed. S. Flader and J. Callicott (Madison: University of Wisconsin Press, 1991), p. 145.
3. D. Brown et al., "Rural Land-Use Trends in the Conterminous United States, 1950–2000," *Ecological Applications* 15 (2005): 1851–1863.
4. W. Berry, "The Whole Horse," in *Citizenship Papers* (Washington, DC: Shoemaker & Hoard, 2003), p. 125.
5. M. Scott, R. Abbitt, and C. Groves, "What Are We Protecting? The United States Conservation Portfolio," *Conservation Biology in Practice* 2 (2000): 18–19.
6. Presentation during the 6th Annual Quivira Coalition conference, January 18–20, 2007, Albuquerque, NM.
7. W. Stegner, "The Sense of Place," in *Where the Bluebird Sings to the Lemonade Springs: Living and Writing in the West* (New York: Random House, 1992), pp. 205–206.

8. Presentation during the 6th Annual Quivira Coalition conference, January 18–20, 2007, Albuquerque, NM.
9. D. Orr, *The Last Refuge: Patriotism, Politics, and the Environment in an Age of Terror* (Washington, DC: Island Press, 2004), p. 106.
10. Ibid, p. 99.
11. Quote on p. 109 in J. Thomas, *A Country in the Mind* (New York: Routledge, 2000).
12. www.quiviracoalition.org
13. A. Leopold, "The Virgin Southwest," in *The River of the Mother of God and Other Essays by Aldo Leopold*, ed. S. Flader and J. Callicott (Madison: University of Wisconsin Press, 1991), p. 180.

CHAPTER TEN

# LAND HEALTH: A LANGUAGE TO DESCRIBE THE COMMON GROUND BENEATH OUR FEET

*Courtney White*

Since my early teens, I have been fascinated with the skin of the earth. It began when my parents rented a dilapidated stable in what was then a remote spot in the desert east of Scottsdale, Arizona. They populated the little stable with an assortment of horses, mostly for trail riding. My favorite was Valentine, my mother's huge quarterhorse, on whose back I began to explore the desert that surrounded the stable like a sandy sea. As I rode, I watched the ground closely—rattlesnakes were Valentine's one and only mortal fear—and as a result my eyes began to see, for the first time, the rocks, the plants, the hills, and the shape of the desert's horizon. Eventually, I developed an appreciation of the land's sandy washes, flat expanses, palo verde forests, and magnificent subtle hues, whose features and colors remain with me to this day.

At fourteen, my interest in the skin of the earth expanded to include dirt. Indulging my budding interest in prehistory, my parents allowed me to join the local amateur archaeological society, which meant that my father spent weekends driving me to a dig in another remote spot in the desert. There, I dug. Under professional supervision, my fellow enthusiasts and I systematically peeled back the layers of a fourteenth-century Hohokam village, ten centimeters at a time. In so doing, I became intimate with dirt.

I kept digging. Over the years I participated in various excavations, including a project in downtown Phoenix directed by archaeologists from Arizona State University. I dug in prehistoric rooms; profiled backhoe-cut trenches; excavated ancient hearths, canals, burials, trash pits; and even helped to uncover a ball court.

My interest in the skin of the earth deepened again when I took the leap from excavation to archaeological surveying—the systematic sweep of land to look for undiscovered sites. For two memorable summers in my late teens, I served on crews whose job was to hike back and forth across the land in straight lines, no matter what the topography, often under a broiling July sun, while scrutinizing the ground for signs of prehistoric activity. I was paid (an extravagant $3.33 an hour), in other words, to be a liminalist—to analyze the fine line between nature and culture. Was a particular alignment of rocks natural or did it indicate the wall of a building? Sometimes it was hard to say, but in the process of asking this type of question hour after hour, day after day, mile after mile, I became sensitive both to the subtleties of the earth's skin and the human imprint on it.

This sensitivity largely explains why I eschewed the usual preoccupations with wildlife and wilderness when I became active in environmental causes during college. I liked birds and animals well enough and had backpacked in countless wilderness areas, but my experience on horseback and on survey taught me that nature started below my feet, with the way water flowed across the land, with the presence of grass and other plants, and with the important way everything combined.

Years later, my experience was confirmed when I met Jim Winder, a rancher near Deming, New Mexico, and Kirk Gadzia, a range expert and educator who lives near Albuquerque. Both employed a language to describe the land that resonated with my youthful exposure to nature. They talked about "land health," employing terms like *water cycle*, *mineral cycle*, and the original solar energy—*photosynthesis*. Jim liked to talk about termites and how they assisted the cycling of nutrients in the soil, a cycling that is critical to healthy plants and, ultimately, the health of the land. Curious, I kept listening.

Then during a workshop led by Kirk in 1998, my interest in the skin of the earth and my curiosity about this new language fused. Kirk took us to the Black Ranch, in the high desert west of Albuquerque (now gone to subdivision), where he pointed to two adjacent patches of ground—one grazed recently by cattle and one ungrazed—and asked, not so rhetorically, which was healthier from an ecological perspective? As an environmentalist I should have known the answer, right? I hadn't a clue. But in my cluelessness I suddenly saw an opportunity. We could start over—from the ground up.

*******

This chapter lays a foundation for dialogue among ranchers, foresters, conservationists, land managers, students, and members of the public at large about stewardship of the nation's open lands, both private and public, based on the concept and vocabulary of land health. Much of the debate over the role of livestock and wood products on these lands, especially the nation's public lands, has been largely a struggle between competing *value systems*, rather than an emphasis on the capacity of healthy lands to produce resources that we use every day in our lives.

Mixed in with this cultural debate is an emerging question among resource professionals, especially federal land managers: What ecological condition were the nation's lands actually in? The answer, most professionals seemed to feel, was poor to middling condition, though no one really knew for sure—and this uncertainty fed the increasingly acrimonious argument between ranchers and foresters and conservationists and environmentalists at the time. A primary reason for the uncertainty was the lack of commonly held standards for assessing land-health conditions based on ecological *function* rather than on what land uses are appropriate for productive lands.

This situation changed in the mid-1990s with the emergence of the concept of rangeland and forest health—which included definitions, attributes, and indicators of land function developed by the consensus of a broad spectrum of forest and range scientists. Their idea of health was a simple but powerful one: For land to produce commodities and satisfy values over time sustainably, it must be functioning properly at a basic ecological level—at the level of soil, plants, and water. And if it is not functioning properly, how we value and use the land will ultimately be in jeopardy.

## Aldo Leopold and Land Health

The term *land health* was coined in the 1930s by the great conservationist Aldo Leopold.[1] He was referring to the ecological processes that perpetuate life—the processes of regeneration and self-renewal that ensure fertility among communities of plants and animals, including the proper cycling of water and nutrients in the soil. Metaphorically, he sometimes likened land health to a self-perpetuating engine or organism whose parts—soil, water, plants, animals, and other elements of the ecosystem—when unimpaired and functioning smoothly would endlessly renew themselves. He frequently employed words such as *stability*, *integrity*, and *order* to describe this "land mechanism," drawing an image of nature that when healthy operated smoothly and ran in top shape.

By contrast, land became "sick" when its basic parts fell into disorder or broke down. This wasn't just a scientific theory. Leopold began to recognize signs of land illness almost from the start of his career as a U.S. Forest Service ranger in 1909. They included abnormal rates of soil erosion, loss of plant fertility, excessive floods and droughts, the spread of plant and animal pests, the replacement of useful by useless vegetation, and the endangerment of key animal species. These examples of disorder in the land mechanism, whether caused by natural catastrophe or by human interference, often led to adverse consequences for wildlife and human populations alike. That's because when nature's ability to regenerate itself over time is damaged—what Leopold called the "derangement" of nature's health—its ability to provide plants for wildlife or food for humans breaks down, as well.

To make his point, Leopold employed another metaphor. Land is like a bank account: "If you draw more than the interest, the principal dwindles."[2] Continue to draw on the interest without building the principal, and bankruptcy is inevitable.

During his lifetime, Leopold observed what happened to human communities when nature's principal dwindled to the point of ruin, the most dramatic, and tragic, example of which was the Dust Bowl. The plowing up of the shortgrass prairie topsoil by tractor for wheat production in the 1920s, followed by a severe drought in the early 1930s, created a disordering of natural and human communities on a vast scale, one that historian Donald Worster called "the most severe environmental catastrophe in the entire history of the white man on this continent."[3]

Over the course of his life, Leopold saw many other examples of the ecological and economic costs of nature's derangement. During his work for the Forest Service in the Southwest during the 1920s, for instance, he drew the connection between overgrazing by livestock in the region's arid landscapes and the widespread evidence of soil erosion, particularly in riparian areas. The loss of plant cover due to overgrazing, he observed, exposed the soil to the erosive power of wind and rain, which quickly resulted in the "disordering" of fragile communities of plants and animals. The human cost of this disordering, he noted, ranged from lost forage productivity—thus reducing the economic viability of ranchers—to the displacement of human communities.

Leopold's responses to these in-the-field insights about land health and sickness motivated much of his work for the remainder of his life. Expressions included, his belief that the mission of conservation is to achieve "harmony between men and land" by keeping the land mechanism in working order; his economic assertion that "healthy land is the only permanently profitable land"; his belief that wildlife populations could be restored with the same tools that had damaged them—ax, cow, plow, fire, and gun—by employing these tools with land health objectives; his demonstration that landscapes could be "read" and understood by their signs; and, ultimately, the formulation of his "land ethic" thesis, in which he argued for a cultural disposition that included people in the "land community" and that protected, restored, and maintained the land's health.

To Leopold, these were all parts of one challenge.

"The true problem of agriculture, and all other land-use, is to achieve both utility and beauty, and thus permanence," Leopold wrote in an essay titled "The Land Health Concept and Conservation." "A farmer has the same obligation to help, within reason, to preserve the biotic integrity of his community as he has, within reason, to preserve the culture which rests on it. As a member of the community, he is the ultimate beneficiary of both."[4]

**Land Health in Today's Conservation**

It is testament to the originality and depth of Leopold's thinking that the land health idea has undergone only two serious elaborations in the ensuing decades. The first revision took place as a result of the rapid expansion of ecology as a scientific discipline after World War II. Leopold's orderly view of the natural world—the engine and body metaphors with their regular arrangement of parts working harmoniously—was replaced with a dynamic, even chaotic, vision of nature as ceaselessly changing, subject to bouts of disruption and stress. This revised idea of ecological health still focused on self-renewal and self-organization, but now scientists see nature as fluid, not static; complex, not reductionistic.

For example, Dr. Bryan Norton, a distinguished philosophy of science professor at Georgia Tech University, has identified five axioms of the natural world:

- Nature is more profoundly a set of processes than a collection of objects; all is in flux.
- All processes are related to all other processes.
- Processes are not related equally but unfold in systems within systems.
- The processes of nature are self-organizing, and all other forms of creativity depend on them; and that vehicle of creativity is energy flowing through systems that generates complexity of organization through repetition and duplication.
- Ecological systems vary in the extent to which they can absorb and equilibrate human-caused disruptions in their creative processes.[5]

In other words, the midcentury view of the "balance of nature" (which Leopold never embraced) was replaced by a view of the "flux of nature," requiring a new set of terms and concepts, including *resilience*, *historic range of variability*, *sustainability*, *diversity*, and *stress*. Moreover, this new vision cast human impact on ecological processes in a new light. Rather than simply upsetting the balance of nature, our activities could now be evaluated according to their roles in the processes of stress, adaptation, and recovery. Those activities that encouraged resilience, for example, could be considered to be promoting land health, while those activities that reduced an ecosystem's ability to recover from a disturbance could be considered deleterious.

This paradigmatic shift among ecologists, however, did not reduce health as a useful metaphor, or as an important goal. "Health is a noun and may therefore suggest a static condition in both organisms and ecosystems," wrote environmental philosopher J. Baird Callicott. "But health, despite the grammar of its name, actually is very much a process, a process of self-maintenance and self-generation. Today, ecologists emphasize that ecosystems change over time, but, like healthy organisms, healthy ecosystems maintain a certain continuity and order in the midst of change."[6]

The second elaboration to Leopold's thinking was a filling in of the specific details that constitute land health. One particular effort began in 1994 with the publication by the National Research Council of *Rangeland Health: New Methods to Classify, Inventory, and Monitor Rangelands* (a similar movement is under way regarding forest health.[7] This effort was a response to persistent disagreement among range scientists, environmentalists, ranchers, and public agency personnel about the health of the nation's 770 million acres of rangelands. Not only was there a substantial lack of data on the condition of the land itself, but there was also an important lack of agreement among range experts on how and what to monitor. These voids contributed significantly to the acrimonious debate raging at the time about livestock grazing on the nation's public lands. Were rangelands improving or degrading? Everyone had an opinion, which was precisely the problem.

In an attempt to resolve this situation, the authors of *Rangeland Health* provided a definition of health: "Health is the degree to which the integrity of the soil and the ecological processes of . . . ecosystems are sustained."[8] Echoing Leopold, they used the word *health* to indicate a condition in which ecological processes are functioning properly to maintain the structure, organization, and activity of the system over time. By *integrity* they meant vigorous energy flows, plant community dynamics, intact soil profiles, and the proper cycling of nutrients and water.

Significantly, they concluded that a "healthy rangeland has the sustained capacity to satisfy values and produce commodities."[9] This was important because it meant that the natural processes that sustain wildlife habitat, biological diversity, and functioning watersheds are the same processes that make land productive for uses such as recreation and grazing by livestock. In other words, when the land is healthy, the ecosystem services we receive from open lands can be sustained.

In keeping with the flux-of-nature theory of ecology, the authors linked health to resilience, which they defined as the capacity of land to recover from perturbations, including floods, droughts, and overgrazing. "The integrity of the soil and ecological processes," they wrote, "determines the vegetation, habitat, aesthetics, and other commodities and values that rangelands can provide and determines how well rangelands are able to resist the destructive effects of mismanagement or natural disturbances."[10]

They also tackled the idea of land "sickness." Physical degradation, they observed, results in the deterioration of the physical properties of soils through compaction, wind or water erosion, deposition of sediments, and loss of soil structure. Biological degradation occurs when there is a reduction in the organic matter content of the soil, a decline in the amount of carbon stored as biomass, or a decrease in the activity and diversity of the organisms living in the soil. "Soil degradation," they concluded, "primarily through accelerated wind and water erosion, causes the direct and often irreversible loss of rangeland health."[11]

Their summary: A healthy ecosystem is one in which erosion is not accelerating, most precipitation infiltrates the soil and is used on site for plant growth or flows eventually to underground storage, the plant community effectively takes advantage of the mineral nutrients and energy that occur on the site, plant composition is dynamic, and ecological functions can recover from natural or human-caused stress.

## Land Health as Common Ground

Following the publication of *Rangeland Health*, a collaborative effort was launched by an interagency team of scientists to develop both qualitative and quantitative criteria for assessing and measuring the health of the land.

A significant step was accomplished in 2000 with the publication of *Interpreting Indicators of Rangeland Health*, which identified seventeen indicators of land health, grouped into three categories:

- *Soil stability.* The capacity of a site to limit redistribution and loss of soil resources (including nutrients and organic matter) by wind and water. It is a measurement of soil movement.
- *Watershed function.* The capacity of the site to capture, store, and safely release water from rainfall and snowmelt; to resist reduction in this capacity; and to recover this capacity following degradation. It is a measurement of plant-soil-water relationships.
- *Biotic integrity.* The capacity of a site to support characteristic functional and structural communities in the context of normal variability, to resist the loss of this function and structure due to a disturbance, and to recover from such disturbance. It is a measurement of vegetative health.[12]

There is an important caveat to these definitions and terms, however: Land health does not necessarily indicate the full realization of a land ethic. A brown trout or an exotic weed might be perfectly *functional* in a particular landscape—filling an ecological niche in a stream or holding the soil together—but it may be unacceptable to a landowner, agency, or community from the perspective of sustaining cultural or biological diversity.

These indicators, terms, and definitions (table 10.1) form a common language to describe the common ground beneath our feet. This language not only creates a solid foundation for dialogue about the future of the nation's open lands, it also provides a means to tackle "land illiteracy"—a quiet epidemic with profound consequences—among citizens of all stripes. Like any illiteracy, our modern-day inability to "read the land" consigns us to ignorance. Or as Wendell Berry has written, "Until we understand what the land is, we are at odds with everything we touch."[13]

| **Table 10.1**<br>**A Partial Land Health Glossary** |
|---|
| **Annual production.** The conversion of solar energy to chemical energy through the process of photosynthesis. It is represented by the total amount of organic material produced within a certain period of time. |
| **Bare ground.** Exposed soil that is susceptible to raindrop splash erosion—the initial form of most water-related erosion. It is the opposite of ground (vegetative) cover. It is vulnerable to capping (soil crusting). |
| **Biomass.** The total amount of living plants above and below ground in an area at a given time. |
| **Cool-season plant.** A plant that grows generally during the early spring, late fall, and winter. |
| **Ecological processes.** These include the water cycle (the capture, storage, and redistribution of precipitation), energy flow (conversion of sunlight to plant and animal matter), and the nutrient cycle (such as nitrogen and phosphorus cycling through the biotic components of the environment). Ecological processes functioning within a normal range of variation will support specific plant and animal communities. |
| **Functional integrity.** The capacity of a site to support characteristic functional and structural communities (soil and vegetation) in the context of normal variability and to resist the loss of that function by disturbance. |
| **Headcut.** An abrupt elevation drop in the channel of a gully that accelerates erosion. |
| **Infiltration rate.** How fast water enters the soil. When restricted (by soil crusts, compaction), water does not readily enter the soil; it moves downslope as runoff and eventually evaporates. As a result, less water is stored for plant growth, resulting in less organic matter in the soil, which weakens soil structure and can further decrease the rate of infiltration. |

| |
|---|
| **Landscape function.** How well a landscape captures, stores, and uses scarce resources, including water, minerals, and organic materials. Dysfunctional landscapes lose those resources to runoff and wind erosion. |
| **Litter.** Any dead plant material that is in contact with the soil surface. It provides a major source of the organic material for on-site nutrient cycling. Also, the degree and amount of litter movement is an indicator of the degree of wind and water erosion. |
| **Oxidation.** A chemical process of decomposition whereby nutrients are released into the atmosphere instead of into the soil. |
| **Pedestals and terracettes.** Pedestals are rocks or plants that are elevated as a result of soil loss by wind or water erosion. Terracettes are benches of soil deposition behind obstacles caused by water movement (not wind). |
| **Rills and gullies.** Rills are small erosional rivulets that do not necessarily follow microtopography as normal water flow patterns do. Gullies are channels that have been cut into the soil by moving water. Both are generally caused by accelerated water flow and result in the down-cutting (rapid erosion) of soil. |
| **Soil.** Consists of mineral particles of different sizes (gravel, sand, silt, and clay), organic matter, and numerous species of living organisms. Soil has biological, chemical, and physical properties, some of which change in response to how the soil is managed. |
| **Soil quality.** The capacity of a specific kind of soil to function within natural or managed ecosystem boundaries, sustain wild and domestic plant and animal productivity, maintain or enhance the quality of water and air, and support human health and habitation. Changes in soil quality affect the amount of water from rainfall and snowmelt that is available for plant growth, runoff, water infiltration, and the potential for erosion; the availability of nutrients for plant growth; the conditions needed for germination, seedling establishment, vegetative reproduction, and root growth. |
| **Soil stability.** The ability of soil structures (groups of soil particles) to resist degradation. When organic matter (roots, litter) breaks down over time, it creates a "glue" that holds soil structures together, which is critical for bio- |

| |
|---|
| logical activity, root growth, and water percolation. Conversely, when soil structures become unstable due to disturbances such as raindrops, flowing water, trampling, earth moving, and other activities, structures can break apart, exposing organic material to decomposition and loss. |
| **Threshold.** A transition boundary that an ecosystem crosses, resulting in a new, stable state that is not easily reversed without significant inputs of resources. |
| **Warm-season plant.** A plant that grows mostly during the late spring, summer, and early fall and is usually dormant in winter. |
| **Well-managed land.** Land that has properly functioning ecological processes, biotic integrity, and soil stability associated with its human uses. |

*Note:* Information given here was culled from Society for Range Management publications and M. Pellant, *Interpreting Indicators of Rangeland Health.*

## Conclusion

Sixty years ago, Aldo Leopold defined land health as "the capacity of the land for self-renewal" and described conservation as "our effort to understand and preserve this capacity."[14] As we move deeper into the twenty-first century, his words take on a renewed sense of urgency. If the Dust Bowl was the major alarm bell of Leopold's generation, a worldwide decline in ecosystem services is sounding the alarm for our generation. Both crises share a common concern—the deleterious impact on human well-being caused by environmental degradation. They also have a common root cause: an industrial economy that convinced us that we could "have our cake and eat it too"—that that there would be no long-term limits or consequences to relentless economic growth. Leopold gently cautioned us on this point—but seventy years after the Dust Bowl we seem to have not taken his words to heart.

On the other hand, things might still turn out differently. Fortunately, thanks to the diligent work of many researchers and practitioners, we not only have a clearer picture of what constitutes land health, we also have good, practical, and profitable models of what constitutes sustainable land use. The big job now is to put that knowledge to wider use, and quickly.

**Notes**

1. See A. Leopold, *A Sand County Almanac and Sketches Here and There* (New York: Oxford University Press, 1948); C. Meine, *Aldo Leopold: His Life and Work* (Madison: University of Wisconsin Press, 1988); and J. Lutz Newton, *Aldo Leopold's Odyssey* (Washington, DC: Island Press, 2006).
2. A. Leopold, *For the Health of the Land: Previously Unpublished Essays and Other Writings*, ed. J. Callicott and E. Freyfogle (Washington, DC: Island Press, 1999), p. 162; and Newton, *Aldo Leopold's Odyssey.*
3. D. Worster, *Dust Bowl: The Southern Plains in the 1930s* (New York: Oxford University Press, 1979), p. 24.
4. A. Leopold, *For the Health of the Land,* p. 225; and Newton, *Aldo Leopold's Odyssey.*
5. Bryan G. Norton, "A New Paradigm of Environmental Management," in D. Rapport et al., eds., *Ecosystem Health: Principles and Practice* (Oxford, UK: Blackwell Science, 1998), p. 25.
6. J. Baird Callicott, "Aldo Leopold's Metaphor," in D. Rapport et al., eds., *Ecosystem Health: Principles and Practice* (Oxford, UK: Blackwell Science, 1998), p. 52.
7. See R. Emonds et al., *Forest Health and Protection* (Long Grove, IL: Waveland Press, 2005).
8. National Research Council, *Rangeland Health: New Methods to Classify, Inventory, and Monitor Rangelands* (Washington, DC: National Academy Press, 1994), p. 4.
9. Ibid.
10. Ibid., p. 7.
11. Ibid., p. 8.
12. M. Pellant et al., *Interpreting Indicators of Rangeland Health,* Version 3, Technical Reference No.1734-6 (Denver, CO: U.S. Department of the Interior, Bureau of Land Management, 2000).
13. W. Berry, "A Native Hill," in *The Art of the Commonplace: The Agrarian Essays of Wendell Berry* (Emeryville, CA: Shoemaker & Hoard, 2002), ed. N. Wirzba, p. 27.
14. A. Leopold, "The Land Ethic," in *Sand and County Almanac and Sketches Here and There* (New York: Oxford University Press, 1987), p. 221.

CHAPTER ELEVEN

# RECIPROCITY: TOWARD A NEW RELATIONSHIP

*Peter Forbes*

Greg Brown took aim, slowed his breath, and gently squeezed the trigger. It was clear, 42 degrees, with a light wind—a fine morning in October to be in Glacier Bay. The shot hit the seal in the back of the head, forcing its mouth shut instantly, keeping it from taking in water and sinking in the deep waters around Garforth Island. The preferred weapon for hunting seals is the .22 Hornet, because it's light, accurate, and can humanely kill a relatively small mammal, but Greg and his uncle were not there for sport. They were there to take a seal and bring it back to Hoonah, a Tlingit community outside of the Glacier Bay Park, for a potlatch ceremony honoring Greg's cousin, who had died. Greg, whose Tlingit name is *Shaaa-yakw-nook*, was doing in 1992 exactly what his ancestors had done since time immemorial, gathering seal, eggs, and berries from a land so critical to their survival that they called it their ice box.[1]

John Muir was the first publicist of Glacier Bay, arriving there by canoe with a Presbyterian minister in 1879. Muir was awed by the vast forces at work in that sweeping landscape of mountain, glacier, and water. Being in Glacier Bay made Muir feel fully alive, and he translated his experiences in a series of popular articles sent in installments to the *San Francisco Bulletin* even before he got back to California. Muir's writing led directly to the creation of Glacier Bay National Monument in 1925 and helped to establish the dominant theme of the early conservation movement: Keep safe what you find valuable by removing people and other species that may threaten it. We have a large and inspiring national park system because of those first efforts at conservation. Tourists and Tlingits alike are grateful that Glacier Bay remains today a largely healthy and whole ecosystem. Muir had a powerful vision that served nature well, but his vision was incomplete: He saw the landscape and not the people.

On that first trip to Glacier Bay 125 years ago, as the story goes, Muir purposely rocked the canoe so that his Tlingit guide would be unable to shoot and harvest a deer. Muir wrote his account to make clear his values, but today it seems a sad parable of two people unable to hear each other's stories about their different ways of being in relationship with a place they both needed and loved.

Greg Brown was arrested later that morning in 1992. His rifle and the hair seal were confiscated by park rangers, and he was ordered to appear before a federal magistrate in Juneau on charges of taking a seal in a national park. The Hoonah Tribal Council quickly came to his defense, saying, "We were made criminals for our food."

* * * * *

This is not a chapter about hunting, or about the management of our national parks; it is about the essential purpose of conservation today, which is to understand the role of land, and our relationship to it, in creating a culture of care and attention in our country. To heal the land, as well as to be healed by it, requires of all of us a deeper self-awareness and a willingness to honestly ask these questions: Who do we allow—and not allow—to experience this land and why? Is our current relationship to this place healthy? What about this land, and our relationship to it, might teach us how to live differently today?

**The Land-People Equation**

With a growing human population and appetite felt everywhere on this planet, it is no longer possible to protect land and nature from people. No property boundary will survive a suffering, greedy humanity. Today's conservationists speak of protecting land through "landscape-scale conservation," but how does that bigger approach "save" land from climate change or acid rain or a public that simply no longer cares? When the human response to a park or wildlife refuge is to develop all the land around the "protected" land, what have we achieved? To be meaningful and enduring, the work of conservation must seek not to work solely on a larger scale or with tougher legal statutes, but to also engage the hearts, minds, and everyday choices of diverse people. The massive, vital work of conservation today is to reweave this still spectacular landscape with the human experience, relating land to everyday human choice and life.

Conservationists have been enormously successful in protecting land, marshalling the money and skills to purchase more than 14 million acres of land across the United States in the past decade, but Americans are no closer, by and large, to that land or to the values that the land teaches. Conservation continues to be swept aside by the homogenizing and insulating effects of technology, electronic media, urban sprawl, and a culture of fear that contributes to the divorce between the people and the land. Today, the purpose of land conservation must be to create a balanced, healthy people who carry the land in their hearts, in their skills, and in their concerns.

An unintended result of the early efforts at conservation has been to exclude many Americans. Conservation must now be defined by the full awareness that our past efforts removed people—primarily the rural poor, people of color, and native people—from the land. People have forever asserted their values over other people in politics, economics, and, sadly, conservation. At Yosemite, the Ahwahneechee were first forced out of the valley, then brought back in to the park to change bedsheets, serve Coca-Cola, and dress up as the more recognizable Plains Indians. At Great Smoky Mountains National Park, almost seven thousand rural people were bought out through condemnation, only to have their barns and cabins reassembled in a Mountain Farm Museum, where actors play at hill country life. More recently, to create the Yukon-Charley Rivers National Preserve in Alaska, a one-hundred-year practice of homesteading was stopped and people removed from their land.[2] Here's one result of this exclusion: Frank and Audrey Peterman, advocates of our national park system, could travel through twelve national parks in three months in 1995 and see only two other people of color.[3] What did we lose as a nation and as a people when conservation became a segregated movement?

The result, too, is that dispossessors are damaged along with the dispossessed. No conservationist will ever reach her or his goal without first gaining a broader sense of history and justice and embracing Saint Augustine's wisdom that one should never fight evil as if it is something that arose totally outside oneself. If you are the one something is taken from, it matters little whether the taker is a robber baron, a land speculator, or a conservationist. Today, we must acknowledge this dispossession of native people and others, such as black family farmers, without whom some significant portion of conservation would not have been possible, and recognize that to heal that wrong—and to heal ourselves—requires not guilt but awareness, humility, and the courage to go forward differently.

Our conservation movement has been guided for more than one hundred years by this question: How do we produce a landscape that is worthy of our culture? But when we say "our" culture, who do we include and leave out? The language of conservation is filled with words about preserving, protecting, and saving places because we know deep down that we are fencing someone out. What we should be fencing out is unhealthy behavior, not whole classes or races of people.

Today is the conservation movement's Age of Becoming. We may have started with a landscape-as-museum philosophy and a focus on one set of cultural needs, but the truth today is that we have conserved vast expanses of land that hold the possibility of a return in a whole way, in a manner never achieved before. This is not going back to the land, it is going forward to the land in a new way. Writer and homesteader Hank Lentfer suggests that we need a new relationship to the land at his home ground at Glacier Bay. "Looking at the clear-cut hillsides around Hoonah, I would be reluctant to return title to the Tlingit," he writes. On the other hand, "watching the smoke billow from the cruise ships idling in Glacier Bay while 2,000 tourists snap pictures with disposable cameras I have to question the wisdom of the 'current owners.' "[4]

Let us consider the possibility that Wendell Berry was right when he wrote more than thirty years ago that "we and our country create one another . . . our land passes in and out of our bodies just as our bodies pass in and out of our land . . . therefore, our culture must be of response to our place."[5] Perhaps the motivating question is no longer how we produce a landscape that is worthy of our culture, but how we produce a culture that is worthy of our landscape.

Most conservationists believe that the land heals people, yes, but only a fraction embrace the alternative possibility, that people heal the land. Humans are tuned to relationships. We are healed by our love and our compassion. And one of the most influential relationships in our lives is with the land itself. We can make soil through composting as well as destroy it through overgrazing. Within some of us still are the skills of how to keep the land and ourselves healthy. That ancient knowledge lies in the daily traditions of the Papago Indians, of Rarámuri farmers, and of Inuit hunters and in the modern skills of homesteaders, forest stewards, and organic growers. What kind of new concept of the land might emerge if we could listen more carefully to one another's stories of the land?

## The Extinction of Human Experience

Day by day, the number of Americans with firsthand experience of the land dwindles. This allows us, as a culture, to destroy land, drifting farther from the anchor that has sustained us physically and emotionally for eons. We see the results everywhere: We have a harder time talking with one another, we have more fears, our physical and emotional health diminishes, and we become more easily manipulated. And soon we find to our amazement that we have become a nation addicted to things, a nation that produces more prisoners than farmers and more shopping malls than high schools.

This separation of people and the land can lead to only one place: a society in which it is no longer necessary for human beings to know who they are or where they live. And if no one knows where they live, then anyone with political power will control the land and the people. Barry Lopez tells us that "for as long as our records go back, we have held these two things dear, landscape and memory, each informing us with a different kind of life. The one feeds us figuratively and literally. The other protects us from lies and tyranny."[6] Our experience and memory of the land, arising from scientific knowledge as well as our human senses of touch, taste, and smell, are the knowledge on which a country must ultimately stand. Our relationship and memory of land, therefore, are deeply connected to our sense of patriotism, citizenship, egalitarianism, and fairness and our sense of limits. A healthy relationship to land is the means by which humans generate, re-create, and renew transcendent values such as community, meaning, beauty, love, and the sacred, on which both ethics and morality depend.

The powerfully corrupting force of disconnection has become business as usual in America, and it is no wonder that conservationists have been afraid to confront it directly. The fact that there are now nearly seven billion people on Earth makes any possibility of a healthy human relationship to land a challenge. As population levels increase and technology amplifies our impact, our capacity for destruction increases. But increasingly the land is without intimates, people for whom the land remains alive, who have indispensable, practical knowledge. Our cultural understanding of land has shifted largely from the personal and physical (that of farmers and hunters, for example) to the industrial and recreational. This is fine except in its extreme, where land simply becomes a form of commerce or entertainment, something to be consumed.

Ten years ago, scientist and writer Robert Michael Pyle coined the phrase "extinction of human experience" in his important book *The Thunder Tree*. He wrote, "People who care conserve; people who don't know don't care. What is the extinction of the condor to a child who has never known the wren?"[7]

The extinction of the condor is the slow, unspoken diminishment of ourselves. It is the damage that occurs when a part of our own capacity to think, feel, and understand is lost because the world around us—the world that shapes us—is also lost. We lose the condor, and we lose some of our capacity to be in relationship with anything other than ourselves and our kind. And the child who doesn't know the wren is the child who is afraid of walking to school, who has already begun to feel boundaries surround her. How will our children love and protect what they do not know?

Here's evidence of the boundaries we make: Today, 42 percent of the private land in America is posted No Trespassing.[8] Conservationists, also, show both our love and our fear by what we fence out; nearly 70 percent of land protected by private conservation organizations is posted No Trespassing. In the span of my lifetime that sign has become America's best-known symbol of our disconnection from the land and a common reminder of our fear of one another. Seeing those signs reminds me of the extent to which we have all become children of a broken lineage.

We humans have prospered from our collective memory of the land, a lineage of direct human experience of nature that functioned for 160,000 years and is now largely broken.[9] We are just beginning as a people to understand the consequences of that fracture. Until this isolation from the land, every human culture had specific words to express their fundamental relationship to it. The Nguni of southern Africa speak of *ubuntu*, meaning "connectedness and social responsibility." The mestizos of the northern Mexico and the southwestern United States have *querencia*, which means "the place and source of one's meaning and responsibility." The Russians have *mir*, which means both "land" and "peace." And the Hawaiians have *kuleana*, which means "personal sense of responsibility" and one's "homeland." Sociologists and psychologists have told us for more than a hundred years that the world we create for ourselves, the economic, social, and environmental systems that surround us (or not), gives us the social clues to be our better or worse selves. Conservation and restoration put into our everyday lives the social clues for how to live well and, thus, help us to be our better selves and to foster a culture of respect, forbearance, tolerance, and peace. This is the extraordinary power of conservation: to help create healthy and whole human and natural communities.

Conservationists have vital work to do. One in four Americans will suffer sufficiently from clinical depression to have to go to a hospital at some point in their lives.[10] Wealth has been consolidated: The richest 1 percent of our population controls one-third of the nation's wealth, creating a more dangerous and immoral divide between haves and have-nots. The poverty rate for African Americans and Hispanics is now nearly three times the rate for whites. These are the realities of American life. As conservationists aspire to speak to a broader range of Americans, we must understand that many Americans are waiting first to understand how conservation is helpful in these everyday realities.

**The Walk toward Whole Communities**

With a spirit of humility and grace, we must ask ourselves the difficult questions that beg us to move beyond our memberships to serve a larger humanity. Can we expand on the motivating questions of our movement, from how much land can we protect, how many laws can we pass, and how much money can we raise, to what relationships do we need to be whole again? What is a whole community, and how do we get there together?

Relationships are as fundamental as places and things. Conservationists made a strategic error in assuming that our work is more a legal act than a cultural act, in assuming that we can protect land from people through laws, as opposed to *with people through relationships.*

Laws codify values; they do not create them. If the people in a democracy no longer care about the land, the laws that protect that land will not hold. Imagine, alternatively, how a conservation movement grounded in an ethos of relationship might be different from one grounded in law. In order to protect land we would need to involve as many different people as possible: hunters, biologists, artists, ranchers, loggers, hikers, urban gardeners. We would need a new quality of dialogue and the ability to hear and respect one another's stories and to make mature choices between types of relationships.

Today, who has the right relationship to the land? To know the answer, we will need to initiate inside our organizations and coalitions, and outside among our neighbors, an ongoing dialogue that will ask us to live with considerable tension and uncertainty as we learn from one another. Also, we will need to balance the rational, legal mind-set needed to protect land with the more empathetic, relational mind-set needed to connect people and the land. Let's start with the controversial and difficult work of envisioning a hierarchy of relationships, an understanding that some types of relationships with the land are more important today than others. Resilient relationships, those that have succeeded in place over long periods of time—say, more than five hundred years—deserve our respect. Also, a right and healthy relationship ought to be determined by the quality of use more than by the legality of ownership. Finally, viewed through the lens of relationship, health means balance, mutuality, and resilience. The destruction of other species would be unhealthy because it is ultimately destabilizing. Right relationship might be defined, in part, through the degree that the human is invested in it ethically, morally, and economically and the land is not depleted. Through the lens of relationship, work and livelihood are more valued (as long as the land is not depleted) than recreation because a nation of people living on the land, growing their food and fiber, is more valuable today to the long-term health of the planet than a nation of consumers is. Similarly, a whole community is resilient and endures not just because of the quantity of its protected land but because of the variety and depths of its relationships to all of its land. Finally, our definitions of right relationship must include encouraging people to experiment today by living on the land. Ways of life are best preserved by living them. Museums are critical places to store our knowledge, but they should never replace opportunities for people to continue to evolve on the land. The walk toward whole communities sees the conservation of land as a cultural act to sustain our democratic traditions; to help people become native to a place; to nurture respect and forbearance, independence, and the source of our sustenance.

All people deserve a relationship to the land. The social foundations that enable conservation to happen in this country—namely, the wealth of many of our organizations, the access we have to political and social power, the ability we have to evolve a legal system to our benefit, even our ability to own land and to work effectively with other landowners—reflect a very privileged position. If we use that privilege primarily for ourselves, we ultimately squander the opportunity to create a whole community and diminish ourselves. If we use that privilege to make meaningful relationships with land available to all people, we have taken what was never really ours in the beginning and turned it into something of value for everyone. The core challenge to conservation today is to create trust and dialogue among diverse peoples. A whole community is built on a *moral landscape* in which people treat one another fairly and other species of life are respected.

Last, consider the power of story. We tell stories to cross the borders that separate us from one another and to help us imagine the world—past, present, future—differently. Stories enable us to see through the eyes of other people and open us to the claims of others. Stories help us dwell in time; they teach us empathy and how to be human. Stories are the way we *carry the land inside of us*. Stories of the land awaken and rekindle the experiences of wholeness inside each and every one of us. This is what Wallace Stegner meant when he wrote, "No place is a place until it has had a poet."[11]

Stories help us to listen to one another and to find different renderings of what is valuable. The shades of love that people feel for the land, whether they are new to that place or have been there for generations, are adequately expressed only in terms of human emotion: the expression of our deepest felt values. Telling stories about our values helps conservationists to explain the role that land plays in shaping healthy human lives. When I tell you who I am and you tell me who you are, our isolation as people and leaders comes to an end, the reweaving of our conservation movement begins anew.

The most important work that can be done today is to create safe harbors where different people can have honest and sustained dialogue with one another about the land: its meaning, what we value in it, our vision about it, and our capacity for sharing it. We need places where people can ask reciprocal questions: Why do I need you, and why do you need me? Why does the health of the land and people need us working together? Wayne Howell, of the National Park Service, is investing years in creating a new relationship with the Hoonah people at Glacier Bay by hearing their stories and reconnecting them with that landscape by organizing trips to pick berries, harvest eggs, and perhaps one day even hunt seals again. The focus of each organization in the environmental movement on individual pieces of the drama, making its own arguments to its own audiences, is why collectively we have not been able thus far to offer a compelling new story for how to be an American. It is also why our movement places a much greater emphasis on strategies and tactics than on story. The former are perceived as "hard," and the latter are perceived as "soft." But without both in equal measure our movement can never flourish. Martin Luther King did not say, "I have a plan." He said, "I have a *dream*," and he spoke of his values without offering strategy and tactics about how we might achieve them. He knew that if he could reach people with shared values, he could respect them to move in the right direction of their own accord. Today's "I Have a Dream" speech for conservationists could be a story about children; about a return to healthful, local food; and about healing the isolation and divides between us all. Healthful food and healthy children are today's most important "doorway issues" to enter more Americans' homes with a new story about land, people, and health. Imagine how many millions of Americans would take conservation seriously if its focus were the protection of our children and our food.

**Conclusion**

The first era of conservation was about protecting physical places because that's what was most needed. The language and skills of that era have been technical and legal, and its goals have often been narrowly focused on the land, often at the exclusion of people and community. The second era of conservation, which we have now begun, is about reconnecting people and the land, about helping people to make choices in their lives based on their relationships with the land and with one another. The leadership skills most needed now are cultural competency, the ability to listen to and create dialogues among diverse people, empathy, and the courage to speak about the value of the land to all people and admit past mistakes.

The new breed of conservation leaders run their organizations not like a business but like an ecosystem where every organization has its own specialized niche in which there is both competition and cooperation. Each organization would be dependent on the success of other organizations. Survival of the fittest does not mean survival of the toughest or survival of the one with the best messaging campaign; it means survival of those who best evolve. Today, successful organizations are those that can quickly form new alliances, share resources, pick up new tools, and adapt to changing conditions. Today's fashionable Resilience Theory says that ecosystems work best when there are strong feedback loops helping organizations and the system as a whole to learn through *experience of current conditions.* The new conservation leaders have moved beyond "staying on mission" to lead by responding to what is actually happening in the world right now. They are regularly speaking of their vision for the future, finding the language and story that reach more Americans, recognizing and speaking aloud past mistakes and injustices.

When leaders and their organizations work in this manner, new life flows to them. They become less brittle, more flexible, and better collaborators. They are putting the fragmented pieces of their lives and of our movement back together. These conservation leaders are using their land for food production and buying new land to create permanent locations for farmers' markets. They are processing sustainably harvested wood from conserved land for affordable housing. They are committed to building wealth for people with low incomes by selling their own restricted land to co-ops, and they are translating their newsletters and Web sites into Spanish. They are with local conservation groups committed to making a meaningful response to global issues like climate change and privatization of water.

Future generations will look back at the creation of very different parks, like Glacier Bay in Alaska and Central Park in New York City, with the same gratitude: They remind us of what it means to be human in healthy relationship to the world. We have been right to act quickly and to save these places from the grinding, numbing wheel of the industrial revolution. The vital work today is to reweave people and the land with the specific intention of creating a more resilient community, one that can be achieved not through fencing people out but only through the far more challenging work of inviting people in. We will never replace the dominant culture of fear and emptiness with a culture of care and attention until more Americans, of all colors and classes, carry the land in their hearts and minds.

**Notes**

1. See W. Goldschmidt and T. Haas, eds., *Haa Aani, Our Land: Tlingit and Haida Land Rights and Use* (Seattle: University of Washington Press, 1998).
2. For a complete account of the Yukon-Charley National Preserve, as well as other valuable ideas on the role of homesteading in wilderness, see D. O'Neill, *A Land Gone Lonesome* (New York: Counterpoint, 2006).
3. See A. Peterman, *Hidden in Plain View: A Black Couple Reveal Secrets of Our National Parks, Public Lands and Environment* (Atlanta, GA: Earthwise Productions, 2006).
4. Personal correspondence with H. Lentfer, July 22, 2007.
5. W. Berry, *The Unsettling of America* (San Francisco: Sierra Club Books, 1977), p. 108.
6. B. Lopez, *About This Life* (New York: Knopf, 1998), p. 142.
7. R. M. Pyle, *The Thunder Tree* (New York: Lyons Press, 1998), p. 145.
8. National Private Landowners Survey (Athens, GA: Environmental Resource Assessment Group, 1997).
9. Two of the most helpful books I've read recently on this topic are E. M. Thomas's *The Old Way* (New York: Farrar, Straus, Giroux, 2006) and H. Brody's *The Other Side of Eden* (New York: North Point Press, 2006).
10. P. Whybrow, *American Mania* (New York: Norton, 2005).
11. W. Stegner, *The Sense of Place* (New York: Random House, 1992).

CASE STUDY EIGHT

# COMMUNICATION NETWORKS, LEADERSHIP, AND CONSERVATION IN AN AFRICAN SEASCAPE

*Beatrice I. Crona*

Views on nature conservation are changing, and so are practices—from federal policies to local initiatives, from top-down to bottom-up, from views of ecosystems as having singular steady states to adaptive management, and from a focus on purely biological resources to inclusion of the resource users. This is wonderful news, and it is about time. The concept of a human-dominated environment has never been more real, and, consequently, humans need to be included as an inextricable part of almost all ecosystems. But as we include humans, the management task grows increasingly complex. Why? Because people have varying agendas, wants, needs, and means, and they all vary among subgroups of society—even in small communities.

Adaptive management is about realizing that surprise is inevitable and that continuous learning is a prerequisite for sustainable management. Adaptive comanagement is about including local stakeholders in the process to contribute their knowledge and understanding of how the system works. But when we do that, we need to acknowledge that even within local communities, there may be vast differences in the knowledge possessed and in the trust and willingness to conserve. Drawing upon my work in Africa, and falling back on theory from the field of common property resource institutions and sociology, I will tell you a story of how a seemingly simple study of local ecological knowledge in a Kenyan fishing village led to a much more complex view of the communication of such knowledge among groups in the community and the effect of the communication pattern on consensus about the status of the resource and, ultimately, on the incentives to manage the resource sustainably.

The setting is a rural coastal fishing community of approximately one thousand inhabitants in Kenya (figure C8.1). Most villagers are dependent on the nearby seascape, including the mangroves, seagrass beds, and reefs at the mouth of the bay, where they fish and collect marine and forest products. Gradual weakening of traditional governance structures and a lack of capacity to enforce the top-down regulatory measures imposed by national agencies has left the inshore fisheries along the coast as virtually open-access systems. This has led to overfishing and depletion of inshore stocks. In response to the situation, initiatives toward more inclusive management have recently been taken. For example, a group of local community-based organizations, community leaders, local administrators, and government agencies was formed with a goal of coordinating integrated coastal management in the area, but local involvement was complicated by conflict. At the time of initiation one idea was that flexible and sustainable management could be helped by daily information about the state of the fishery from fishermen who spend long hours actively observing the resource.

With the intent of studying the local ecological knowledge of the fishermen, I interviewed groups of resource users based on how they use the resources. This was done because, after spending considerable

Figure C8.1 Fishmongers at a local landing site waiting for the fishermen to return with the daily catch.

time in the village, I suspected gear would partly dictate fishing area and affect the knowledge accumulated by each group. As it turned out, the truth was not far off. Knowledge did differ among groups of fishermen, partly as a direct consequence of spending most of their time in one specific area and observing the natural system in that spot. Consequently, reef fishermen knew more about the reef than women fishing shrimp in the mangrove fringe, and seine netters operating in the lagoon had understandings of seagrass and urchin interactions not acknowledged by other groups. When one considers that the village had only two hundred households, and most depended on fishing, it was surprising that such differences still existed among the groups. I often saw different types of fishermen chatting over cigarettes outside the shop. What did they discuss then? Didn't they exchange information about fishing?

Intrigued by the differences in knowledge among fishermen, I set out to map the social network of communication of resource-related information in the community by asking all heads of household with whom they exchanged knowledge about their natural environment and information useful for carrying out their occupation. This resulted in a network of over 150 individuals with relational ties to one another, which at first looked very messy. But when I looked at the occupation of each individual, a pattern emerged that showed several subgroups that did not seem to communicate directly with each other (figure C8.2). Some groups (like the seine netters) were peripheral, while others were central (deep-sea fishermen) and enjoyed a higher status in knowledge transfer because people came to them to

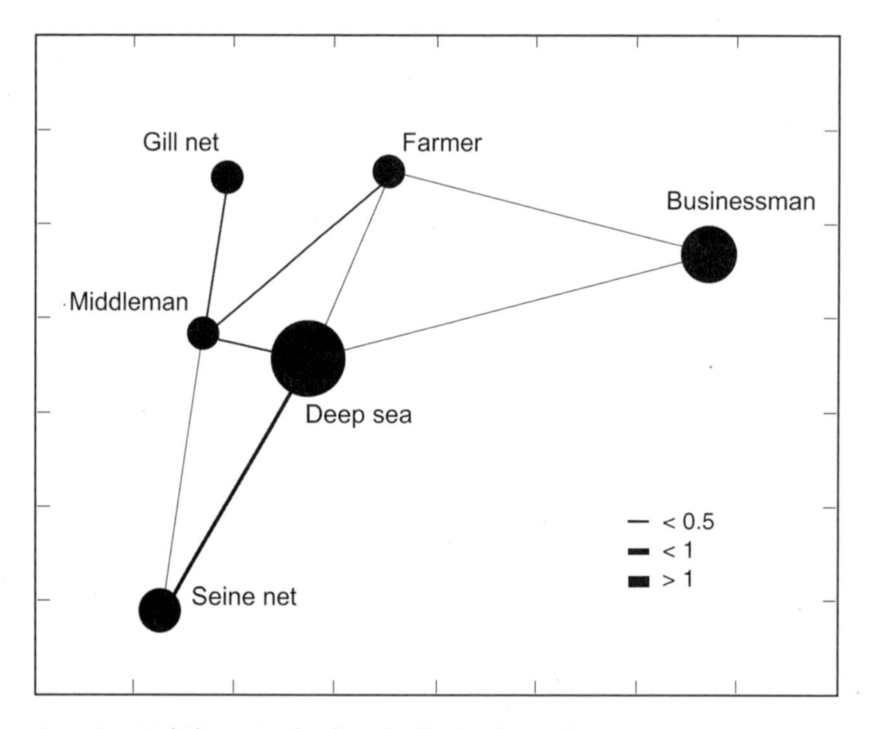


Figure C8.2 Multidimensional scaling plot showing the social network structure of relations among groups based on occupational categories. The size of each node is proportional to the group size, and the thickness of the links is proportional to the strength of the intergroup relations—i.e., the ratio of the observed versus the expected number of relations. Worth noting is that no direct exchange between fisher categories Seine Net and Gill Net occurs. The distance between the nodes indicates how similar each group is in terms of the types of links they have to other groups.

exchange information. The heterogeneous knowledge of the different groups correlated well with the structure of the social network. Groups that were more peripheral and had weaker ties to other groups shared less knowledge with the central groups. Groups of nonfishermen had little resource-related knowledge at all and also weak ties to the fishing groups.

So fishermen were not a homogeneous stakeholder group. In fact, most resource-related communication occurred within distinct subgroups, defined by gear type. Another interesting pattern is that while deep-sea fishermen were more knowledgeable and central in the network, they were also represented by mostly migrant fishermen. Approximately 37 percent of village households were recent immigrants, and the majority of them originated from Tanzania. Many of the Tanzanian fishermen reside in the village on a semipermanent basis, returning to their homeland during seasons of low fishing activity. This seems to have affected their incentives for conservation due to lack of a sense of place and long-term commitment to the area.

Self-organization of stakeholders and collective action have been suggested as important for adaptive comanagement, and diversity of knowledge is thought to enhance the adaptive learning process. Success is also influenced by the ability of parties to agree on resource-related problems and resource status. But fisheries are complex, and with heterogeneous knowledge distribution and groups of fishermen who do not communicate freely among themselves, reaching consensus about the status of a resource is tough. Agreeing on what measures to take and acting collectively may be even harder and often require some form of leadership. But in a village such as this one, who are the leaders to initiate such a process?

At the individual level, the social network had more to tell. Some individuals were much better connected than others. People came to them to exchange information, and they appeared to be informal leaders of opinion. So who were they, and why had they not instigated collective action for the benefit of the declining resource?

The network identified twelve key individuals, including the presiding chairman (elected semiofficially by the villagers) and the subchief, the lowest level of government representation in the village. However, 60 percent were deep-sea fishermen of semimigrant status, the group that had been shown to be less committed to declining fisheries in the area. Further interviews with key individuals revealed that over 80 percent did not even acknowledge that the current situation could jeopardize the continuation of fisheries-based livelihoods. This was likely due to their reduced sense of place and to their higher mobility, roaming deeper waters with larger vessels. Regardless of the reason, however, this was a striking result. If persons centrally positioned in a network of communication dedicated to resource-related information exchange are either unaware of the urgency of a declining resource or do not perceive it to be a problem for themselves, that is bound to have a profound effect on efforts to manage the resource. More precisely, how does that affect the ability of fishermen, as a general group, to reach consensus about resource status and, furthermore, about what measures to take?

This takes us to the final point of this story—that of leadership. Why are leaders or key individuals even important for natural resources management, and why should anyone interested in nature conservation be concerned with their existence at all?

I argue that key individuals who provide leadership and guidance through the complex and turbulent process of organizing collective action are vital for well-functioning comanagement. They facilitate collective action, negotiate between conflicting interests, and help resolve conflict. Naturally, the role and importance of a key individual are related to the skills and attributes of that person. Provided such individuals have the relevant education, resources, social capital, or social networks that extend outside community borders to state and nonstate agencies, they have the ability to assist others in the community with the know-how to maneuver the bureaucracy. This may be vital for securing funds and support when establishing conservation strategies and management plans or developing alternative livelihoods. All of these social and economic factors will fundamentally impact any initiative to conserve and sustainably manage a natural resource on which many people depend for their livelihood.

If we return for a moment to our case—what is the importance of the findings presented here? Why does it matter that the fishing community turned out to be rather heterogeneous, while the opinion leaders in the village were markedly homogeneous? Several points can be made with respect to this. For one, centrally positioned actors in a communication network have a natural advantage in that they receive and potentially control much of the information. From an entrepreneurial perspective this is ideal because new ideas and information can be obtained quickly and responses to change designed early—precisely what an adaptive manager would want. But if the majority of actors in those positions neither perceive a problem nor have incentives to act on it, it is unlikely that they will instigate any collective action for resource management, or even look for information about negative changes in the resource. Their homogeneity in terms of occupation, and consequently knowledge, is also likely to affect their ability to perceive and synthesize different kinds of information. As such, it reduces their ability to adapt to new circumstances (e.g., decline of fish stocks) and potentially lowers the community's adaptive capacity and ability to respond to change and disturbances.

What can be learned from this case? I believe this story contains several lessons that are valuable for nature conservation issues and can be transferred to settings beyond rural Africa. First, even seemingly small and culturally homogeneous communities are likely to contain subgroups with different (and sometimes opposing) views of how the natural system works; whether management is needed; and, if so, what kind. Second, consensus around resource status and how the natural system actually works may

CASE STUDY NINE

# FARMER AS CONSERVATIONIST

*James O. Andrew*

My family has been involved in agriculture in Iowa for five generations. Our ancestors immigrated to the United States in the 1850s after realizing that the farmland of Scotland was worn out and hearing of the rich, black soil to be found in the upper Midwest. Today, we farm three Iowa Century Farms, meaning that each has been continuously in our family for over one hundred years. The earliest land was purchased by my ancestors in 1855 from the U.S. government for $1.25 per acre. When I returned home to farm in 1974, after four years of college and three years active duty in the army, I felt a sincere desire to honor the memory of those ancestors by maintaining and improving the land they had cared for through wars, depression, drought, and other hardships. It was my turn to step up to the plate.

I was blessed with a wonderful visionary mentor and teacher in my father. We felt early on the local pressure to increase the number of acres we farmed, 1,350 at that time; and we did not like the ever upward-spiraling land prices and competition it evoked among neighboring farmers. Therefore, we made a calculated decision that it would be better to improve the health of the land that we owned than to jump into the bidding war for additional acres.

Our first objective was to conduct an honest evaluation of our total resource package and identify shortcomings in addressing our conservation needs. At that time, plowing in the fall was the accepted tillage system for our heavy black soils. We were sickened, however, to see the windblown soil on white snowdrifts in the spring. Thus, we converted to conservation tillage with the use of a chisel plow and field cultivators. That system was a marked improvement over traditional methods, but it left us desiring further change to avoid the water and wind erosion of our soils.

Working with the county Soil Conservation Service (forerunner of the present Natural Resources Conservation Service), we set about designing a series of parallel-tile terraces and outlets, grass waterways, three farm ponds, and other improvements that would control soil erosion on our gently rolling landscape. When that project was completed over a period of ten years, we had installed three and one-half miles of terraces and thought we had done everything we could do to minimize soil loss.

However, the forces of Mother Nature would prove our judgment wrong. We still witnessed sheet erosion from water and dust storms, despite our use of conservation tillage. In 1993, after three years of study and planning, we made a leap of faith, abandoning conservation tillage and shifting to no-till in a strict corn-soybean rotation. In doing that, we sold eighteen tractors, field cultivators, chisel plows, and discs to finance a new corn planter, bean drill, 125-horsepower tractor, and Spra-Coupe self-propelled sprayer. Being among the first in our area to practice no-till farming, we sometimes felt we were farming in a fishbowl, as our neighbors were skeptical of what we were trying.

No-till farming involves no plowing of the soil surface. In other words, we do not disturb the corn stalks or bean stubble after harvest in the fall; we plant the following year's rotational crop into the untouched soil the next spring. This allows the preceding crop's residue to blanket the surface and protect the soil from water and wind erosion while building up organic matter as it decomposes over time. It also turns the fields into giant sponges, as the stalks and stubble soak up the moisture, and, through the work of earthworms, rainfall is allowed to percolate into the soil and soak in. Our no-till fields can hold a five-inch rain before ponding on the surface will occur, while our tillage neighbors have large drowned-out areas. This is also a plus as we enter drought-cycle years when available soil moisture is crucial.

After fourteen years of no-till, we would never consider returning to our tillage habits of the past. In addition to the soil conservation benefits, we enjoy reduced costs of labor, equipment, fuel, and time. These savings allowed our farms to be included for several years in the Top 100 Best Managed Farms in the United States, as evaluated by *Farm Futures* magazine. With today's petroleum prices, we are able to produce the same crops as in the past with about half the fuel needs. The time savings have allowed us to design further improvements and address other conservation concerns such as trees, dikes, and wildlife habitat in our operation.

At our three farm pond locations, we began planting trees and shrubs to provide cover for the wildlife we hoped would inhabit the area. We have installed nesting platforms for Canada geese and are gratified to see that they are occupied in the spring. Birdhouses have been installed for bluebirds, purple martins, wood ducks, and other native species. We also leave unmowed areas around the pond to allow other wildlife to hide and nest. All the ponds are stocked with fish, and native flora abound around the edges of the waters. While my father and I are not avid hunters or fishermen, we visit these beautiful areas to unwind from our daily duties.

Hardin Creek flows through two of our bottom-ground farm fields and during periods of heavy rains has a tendency to overflow its banks and flood our adjoining fields. Long before filter strips, we built earthen dikes along these susceptible areas and installed one-way drain tubes, allowing any water on the field side of the dike to drain into the creek as soon as the rain subsided. This improvement keeps us from farming right up to the creek bank and filters any nutrients before they can enter the water supply. At the same time, it has reclaimed crops from areas previously abandoned because of drowned-out areas.

With heavier and larger farm equipment, we have avoided soil compaction by seeding twenty-foot grass strips on the headlands of most of our fields. We use these areas to refill planters and sprayers and park trucks for hauling away grain. We allow no grain carts or vehicles other than the planter, sprayer, and grain combine on the no-till areas of our fields. This has all but eliminated soil compaction in our operation.

Because we no longer cultivate our crops, we use herbicides to control weeds that arise. One of the additional advantages of no-till is that weed pressure declines over the years since you do not disturb the soil by planting additional weed seeds when you cultivate. We use a dual-action herbicide program of Liberty corn and Roundup Ready soybeans to ensure no herbicide-resistant weed issues arise. We are able to use reduced rates of these herbicides, further benefiting the environment.

In 1997, we dedicated three hundred acres of corn to scientific on-farm nitrogen strip trials conducted by scientists at Iowa State University. That research convinced us to change our application of anhydrous ammonia from fall to spring and to reduce our nitrogen fertilizer application with no yield reduction. The research has proven that we cannot replicate the advertising claims the manufacturers promise and thus help others avoid costly mistakes on their acres.

In 2005, our Raccoon River Watershed was selected for inclusion in the newly authorized Conservation Security Program (CSP), which was designed to reward the best conservation stewards and challenge others to follow. Feeling we were in as good a position to qualify as anyone after some thirty years of planning and preparation, we applied at the local Natural Resources Conservation Service (NRCS) office. We submitted our data and after a thorough evaluation were awarded Tier III status, the highest level, and an annual award of $45,000 for our previous and continuing conservation efforts.

Our selection as one of the first Tier III qualifiers in the United States took on celebrity status in the farm press and media. Because of this, I decided the best use I could make of my newfound expertise was to accept many opportunities to talk with others regarding conservation and lobbying efforts to expand and rejuvenate the CSP program for others. I have been able to speak to over three thousand farmers nationwide, testify before the Senate Agriculture Committee in Washington, and work with the Iowa and American Soybean associations and the Sand County Foundation on new programs to address our nation's soil, water, and air-quality challenges. Presently, the Iowa Soybean Association and the Sand County Foundation are designing a water quality evaluation program for the Upper Mississippi River basin to reduce the soil and nutrients flowing to the Gulf of Mexico and causing increased levels of hypoxia. We have been in Washington, lobbying Congress and the NRCS for needed funds to design and implement this program in Wisconsin, Iowa, Minnesota, Illinois, Missouri, and Indiana. This has also allowed several crucial alliances to develop between the agriculture and environmental communities and, one hopes, will lead to win-win solutions for both parties. Thus, the farmer-conservationist job is never done, as we attempt to be proactive and solve important conservation problems facing our region.

My years of working with conservation projects on and off our farm have proved personally rewarding. Beyond the financial rewards is the satisfaction that we are attempting to ensure that future generations will assume responsibility for our land in better condition than when we started farming it. This is our stewardship vision. Experience has proven to me that the incentive reward program is far more palatable to the working farmer than regulation and mandate. As I witness more and more conservation planning being done via satellite maps and computers, I prefer to take time to return to the land and visually sense the right thing to do rather than follow a prescribed solution from a manual written in the confines of an office. In spite of our zealous desire to immediately correct the sins of the past, it is important to develop patience and realize that many of these problems cannot be corrected overnight. We should design programs that will address conservation issues over the long haul rather than always look for the quick fix. Sometimes the best way to go fast is to begin by going slow. On our farms we have learned how to farm better by farming adaptively. With over thirty years of planning, application, and monitoring, that is one thing, at least, that I am convinced of.

CASE STUDY TEN

# WALLOWA COUNTY: THE POWER OF "WE"

*Diane Daggett Snyder*

The sun peeked through the openings in the ponderosa pine forest, providing a natural spotlight on the two men who stood talking to a group of onlookers. They poked fun at each other occasionally, as people who work together often do. These two men exemplified the depth of collaboration and community leadership in public forest land management in Oregon's Wallowa County; one represented an industrial forest landowner, and the other represented the most active environmental organization in the area (figure C10.1). The subject of their talk was the marriage of forest management and conservation. They spoke of the community-led watershed assessment, in which core values and shared learning were the drivers for a noncontroversial approach to public forest land management. Those in attendance stood among the trees and witnessed a new era unfolding, an era that is operating through partnerships and collaboration, empowering people to be involved in decision-making processes, and encouraging grounded conversations about critical issues facing the social and economic health of forest-reliant communities, as well as the contribution of ecosystem health to those communities.

Figure C10.1 Finding common ground in the forest (photo by Ellen Morris-Bishop).

Wallowa County's historic dependence on the timber industry for its economic well-being is similar to the situation of many communities across the West. With a preponderance of public land (nearly 58 percent), vast wilderness areas, and the presence of endangered species, this beautiful corner of the world was ripe for the controversy sparked by an endangered species listing, environmental appeals and litigation, and the technological advances of industry. Over the course of a short six years, timber harvested from public land plummeted from more than 80 million board feet (in 1988) to nearly zero (in 1994). The fear and frustration of community residents were deeply felt and played out in some unfortunate ways. In 1995, a noose was hung out for the forest supervisor of the Wallowa-Whitman National Forest, and two environmental representatives were burned in effigy on Main Street. The loss of jobs was vast, with little apparent opportunity for local residents to affect it. Their futures were hanging in the balance.

Not only was the community affected, but the land managed by the U.S. Forest Service was devastated. Over time, the public land managers have faced enormous increases in the amount of paperwork required to perform a given project that would promote the health of the land. The listing of endangered species and the costs of defending appeals and litigation have taken their toll on agency personnel, as have ongoing budget reductions within the agency, which result in fewer people to do the work. Nevertheless, the bar continues to be raised, as the agency has moved into a conflict-averse position of decision making, investing in projects that have the most opportunity to succeed (the low-hanging fruit), rather than those that will benefit the health of the land the most. And, even more disconcerting is the fact that nearly half the entire agency budget is spent on fire suppression, rather than using limited program dollars for ecosystem health and its contribution to communities in and adjacent to the public lands.

New leadership in Wallowa County eventually changed the trajectory from the conflict-ridden "us against them" mentality to one of collaboration. This leadership brought all stakeholders of public land management together and stimulated the creation of a salmon-habitat recovery plan. Community conversations focused on a new way of doing business—looking toward sustainable resource management with social and economic contributions to the community.

Ultimately, these conversations resulted in an increased sense of ownership of the decisions being made about public land management and a unique understanding that all stakeholders need to be involved for long-lasting solutions to be generated, implemented, and monitored over time.

The work in Wallowa County has generated many benefits, including the creation of a nonprofit, Wallowa Resources (WR), which provides leadership in facilitating collaborative processes and implementing collaborative projects. The organization has also brokered trust. An attitudinal shift is occurring as the work progresses, a new sense of common purpose founded on core values. The stakeholders working together not only see the results of their collective vision and work, but also learn to respect the views of their historic foes.

Over the past ten years, WR has received support from private foundations and state and federal government agencies to implement collaborative projects on public lands. Indeed, for the past six years, WR has invested more in public land restoration projects than the government agencies. The work spans an array of projects, including treating invasive weeds; reducing forest fuels and performing thinning; restoring fish passages; building off-stream water sources for livestock and wildlife; invigorating aspen stands; and performing forest, range, riparian, road, and recreation assessments. In addition, a small-diameter-wood processing facility provides jobs in value-added wood manufacturing, producing posts, poles, and bundled firewood products. Presently, these projects employ the equivalent of nearly 2 percent of the nonfarm workforce in the county.

With Wallowa Resources as the facilitator, the county has embarked on a system of community-led watershed assessments. The Joseph Creek Watershed Assessment now spans more than 352,000 acres. A tiered approach was taken, with assessment occurring on the first half (Upper Joseph Creek) and project prioritization and implementation following. Then the second half (Lower Joseph Creek) began. The assessment is in varying stages of completion in Lower Joseph Creek. Already more than $600,000 has been invested in Upper Joseph Creek for assessment and project implementation.

The stakeholders in Wallowa County have invested their time and resources in the process and have had many heated debates. In the end, this process has proven to be replicable and interesting to other communities. Representatives of the Wallowa County Natural Resource Advisory Committee, who oversee the process for the county commissioners, are regularly asked for presentations and community visits, as is their partner, Wallowa Resources.

This collaborative group not only has been working on the ground, but also has worked hard on policy issues that constrain its agency partners. The first Stewardship Contracting Training was held in Wallowa County in January 1999. This groundbreaking event stimulated the testing of "new tools for new times." These new tools included innovative contracting mechanisms unfamiliar to traditional loggers, who were concerned about negative impacts to their business of serving as the "guinea pig." Contracting officers were unfamiliar with the new mechanism, and contractors were unaccustomed to integrating multiple tasks into a service contract (rather than the typical timber sale contract). After the kinks were worked out, the first stewardship contract proved to be beneficial for the contractor and the public land manager.

The collaborative also worked on the policy language that created the National Fire Plan, ensuring that the phrase "community benefit" be included, recognizing the importance of structuring the work at a scale that can be accessed by the local contractors and help to ensure that local communities will benefit from the outcomes. This national investment under the National Fire Plan helped the collaborative create contractor capacity in the community to perform fuels reduction and thinning projects. In the past these projects were taken by contractors from outside the community. Everyone benefits from actively preparing the landscape for fire rather than spending a majority of available funds suppressing fire.

This is the kind of success story that can be told only when people are committed to working together, accepting the differences that traditionally served to divide them and being curious about those differences. This is the story of one community, but it can be the story of hundreds of communities, given the groundswell of optimism and desire being expressed throughout the country. If we collectively make conscious choices now, the future of public land management will look different. People will be empowered to be part of the decision-making process, and the health of the land and the community will benefit.

Despite all this progress, there are still constraints to collaboration, restoration, and community benefits. In sum, they are the national funding priorities of the Forest Service, continuing litigation, and the attitudes of federal employees who have been in the midst of controversy for a very long time.

Funding for all land management agencies has been declining for the past decade. For the Forest Service, the decline has had a direct impact on the health of the lands they are charged to manage, as well as the communities on and around those lands. As funds decline, most budget reductions have occurred at the ground level. In the late 1980s, for example, there were in excess of 130 people working for the Forest Service in Wallowa County. There are currently 37, yet they manage the same amount of land and comply with increased regulations. To make matters worse, the "color of money" doesn't always align with the local priorities of the district rangers: Money is allocated by type of work and, many times, is restricted for particular uses. For example, despite wildlife habitat or recreation being the top priority for a district, funds allocated from the Washington and Regional offices of the forest service may arrive earmarked for fuels reduction and riparian area restoration instead. Finally, because of the increase in forest fires over the past several years, society is compelled to favor fire suppression activities, so that's where the agency is focusing its resources. Close to half of the agency's budget went to suppression last year, which further constrains funding for other critical public land needs. The process of budget allocation needs to be changed. Distribution of funds should be focused at the district ranger level, rather than the current distribution that originates far from the landscape to be managed.

Public lands are managed by the government in trust for all Americans. Many Americans still express their interest by appealing federal land management actions. The appeals process needs to respect that right, and those who appeal need to be held to a standard that requires their participation and declaration of substance. Forest Service employees need support for their actions. They should be encouraged to be innovative and collaborative. The fear of a lawsuit often results in no action or action that supports less controversial projects.

Come back to the forest with me. Can you see the people as they stroll through pinecones strewn haphazardly across the ground, hear the snap of an occasional twig, remember the faces of those who have come before, and dream of those who will follow? The time is now. We must all walk together, sharing our knowledge, showing respect for the concerns of others and love for the things we all hold dear.

CASE STUDY ELEVEN

# COLLABORATION AS TEACHER

*Dan Dagget*

In the mid-1980s, Arizona was the setting for a perfect storm of environmental controversies. A rush of uranium mining in the side canyons of the Grand Canyon just outside the boundaries of Grand Canyon National Park had sparked resistance from environmentalists and anti-nuke activists. The noise and other impacts created by a booming business in tourist overflights of the canyon had become a national issue. A push was on to change "the stock-killer law" that gave Arizona ranchers all but free rein to kill mountain lions and black bears they considered a threat to their livestock, and a move to reintroduce wolves into the state was well under way. Last but not least, the dispute over the damming of Glen Canyon was as contentious as ever, fanned by the flames of Ed Abbey's rhetoric.

This caldron of controversies became hot enough to give rise to Earth First!, at the time one of the most radical of environmental groups. It also served as the crucible that forged my ideas about collaboration.

I was involved in all of the above issues and via that involvement helped plan some of the first direct actions of Earth First! Among those actions was one in which the main road in Grand Canyon National Park was blockaded, and Earth First! founder Dave Foreman and a number of others dressed in radiation suits were arrested. As part of my involvement in the stock-killer controversies, I helped carry a huge spike-toothed bear trap into the office of a legislator who supported the ranchers; we set and sprung the trap on his desk as cameras rolled. For my efforts I was designated one of the top one hundred grassroots activists in America for the Sierra Club's John Muir Centennial in 1992.

While I was involved in those actions, I can't say that I was dissatisfied by the confrontationist approach I was using or that I was looking for a different way to address the issues. I was enjoying myself.

It was my involvement with the stock-killer law that changed all that. The beginnings of that transformation took place at a meeting in a carport in Phoenix in 1989.

The reason for the meeting in Phoenix was that the battle over the stock-killer law had become so heated and contentious that one of the environmentalists involved, a woman named Bobbie Holaday, had become alarmed that it would jeopardize her life's goal—the reintroduction of wolves into the Southwest. In light of that concern, she and a friend from a ranching family—a woman named Tommie Martin—decided to bring the players together, face to face, to see if the heat of battle could be lowered. In all, seven environmentalists and eighty ranchers received invitations. Six of each showed up.

Tommie Martin was a member of the third generation of her family to live on the 7A Ranch in central Arizona. She was trained and experienced in dealing with contentious meetings and had a reputation for being able to bring confrontational groups to collaboration through effective facilitation.

To me, "collaboration" and "facilitation" were just words. As I took my seat in Bobbie Holaday's carport, my intent was the same as it was at all other meetings—to bash the other side and do everything I could to "win" the inevitable confrontation. I was sure everyone else there was prepared to do the same.

Tommie dispelled this preconception by saying she didn't drive halfway across the state to watch a bunch of enviros and ranchers call each other the same old names. She then asked that, to keep that from happening, we consent to a few simple rules. These agreements were new to me, but now I realize the first few were boilerplate elements of facilitation technique.

She asked that we treat each other with respect. She asked that we all agree to deal with problems within the group rather than take them home as evidence that what was happening here wasn't working. And she asked that we agree to speak only for ourselves and not as representatives of any group or in terms of any disembodied "they."

The next agreements she asked of us were more of a stretch. They were also, I have come to learn, what made her approach unique.

She asked that we commit to make no compromises. I remember being puzzled by that, even as Tommie explained that what she meant was that if we came to any sort of agreement, we would make it according to the principle of what she called "win-win or no deal." I had never heard that phrase, and I didn't know at the time that it came from Stephen Covey's *Seven Habits of Highly Effective People*, but I felt I understood it in some intuitive way and agreed to it. I had no desire to compromise, and I did intend to win.

Last, she asked that we agree that whatever we did at this meeting we would do in order to achieve our own goals. No being nice. No throwing someone else a bone. "If it doesn't get you what you want," she said, "don't do it."

Now I was totally baffled. If we weren't going to give up some of what we wanted, if we weren't going to compromise, it seemed to me that we were guaranteeing that this would be a very short and unproductive meeting. As far as I was concerned, the only way two groups so far apart could work together was to compromise. Because I was at this meeting more out of curiosity than anything else, I agreed to see what would happen.

After a short break, Tommie went right to the heart of her approach. She asked us what we wanted those rangelands to look like? What did we want their condition to be? And what did we want for their future? She told us not to go for some lowest common denominator, not to pick something because we thought it was doable or the others might agree to it, but to aim high. We were listing goals—what we wanted the land and our life in relationship to it to be—not what would get us by without a fight. Some of us held up our hands and began talking. Tommie wrote what we said on the obligatory newsprint tablet and taped the lists on the walls of the carport.

As this part of the meeting unfolded, something very surprising happened—it certainly surprised me. Ranchers started saying things I had planned to say, things that I thought would come only from an environmentalist, a radical one at that: more open space, less people, more wildlands, healthy populations of wildlife, restored streams with water running in them, and ways for people to live in the West that didn't destroy the very reason most of us had moved there in the first place. One rancher even said, "Healthy populations of predators," while another rancher joked, "Maybe not overly healthy." With each additional environmental goal brought up by a rancher, I found myself more and more astonished that my goals and the goals of people I had considered to be my complete antithesis weren't that different. In some cases, they were exactly the same.

When one of us would say something that could be construed as telling those from the other side what to do, Tommie would gently but expertly steer us from posing prescriptive goals to talking about descriptive goals. This was her way of avoiding the confrontation that is almost a reflex when anyone tells anyone else what to do. When one of us would say something like, "I want all the cows off public lands!" or "I want all the fences down!" before there was a retort, she would ask us what we would like to achieve by doing that. She would ask us how we would expect the land to change. How would it look? How would it feel? How would it make us feel? And she would keep asking questions like that until we answered in terms of ends instead of means. The result, in almost every case, was that even goals that initially threatened to divide us ended up bringing us together.

After the goal-listing session, we realized that we really did have much in common—we all wanted many of the same things. Based on this, Tommie pointed out that in a very significant way, if one of us was to lose, we would all lose, and the only way any of us could win was for all of us to win. At that point we realized that she was right, and a feeling of community hit us like someone turning on a light. People smiled and joked. We agreed to meet again and agreed to call our group 6-6 for six of us and six of them. And we continued to meet for several years.

My eyes having been opened by this introduction to collaboration, I went on the road and attended meetings of a number of groups similar to 6-6. As I visited those groups, one thing that impressed me as much as how they were bringing people together was the success they were having on the ground. I was so impressed with what I learned that I wrote a book titled *Beyond the Rangeland Conflict: Toward a West That Works* and took my stories about the collaborative approach on the road. And in every case the response I received was overwhelmingly positive and enthusiastic. What nay-saying I did encounter came from the "captains of conflict," the group leaders who have made a business of this hundred-year-old war between environmentalists and ranchers and have something to lose if the rest of us declare peace.

More than twenty years have passed since the creation of 6-6, and although I'm aware that there have been some notable successes in creating collaboration, what is amazing to me is that so little has been done with so powerful and benign a tool.

The reason collaboration hasn't caught on any better than it has, I believe, is because its practitioners fail to recognize (or fail to sustain the recognition of) the two kinds of goals—prescriptive and descriptive—that Tommie Martin made us aware of at that first 6-6 meeting. That failure can be fatal to any effort to bring diverse people together in a functional group.

Consider any war or movement that is going on in the world today. In fact, consider just about any kind of conflict. Almost invariably it is about a prescription: the right way to worship, help the poor, run an economy, deal with the oil crisis—you name it. Bring any group of people together and try to decide how to do anything—in other words, try to agree on a prescription—and invariably you've got an argument or worse. The groups don't even have to be diverse. People in the same church, same political party, same environmental group get into terrible battles over the best way to do something—anything.

Now consider that every day the immense diversity of people that make up the world's population work together to achieve all sorts of descriptive goals. They make widgets, grow food, conduct commerce. While you couldn't get a collection of Republicans and Democrats; gays and straights; Muslims, Christians, and none-of-the-aboves to agree on who ought to run the country and how they ought to do it, you could get them to build a house. I've worked with groups at least that diverse that enthusiastically have worked together (and succeeded) in restoring a mine site, creating habitat that supports populations of endangered species, and other equally amazing things. In fact, set a goal, just about any kind of goal, and humans will accept it as a challenge and work to achieve it if the result is something that even remotely matters to them.

The advantage in dealing with environmental issues is that healthy environments matter to just about all of us.

I came away from 6-6 convinced that collaboration can work if those participating recognize the distinction between prescriptive and descriptive goals and base their collaboration on the latter. My experience in 6-6 also made me aware of how very, very difficult this can be. On one 6-6 field trip a self-identified eco-radical told a rancher, "There's only one thing you can do to make this place better. You can leave; because if you stay, no matter how good you make it, it will be unnatural and therefore bad. And if you leave, whatever happens, even if this land becomes as bare as a parking lot, will be natural and therefore good."

I know of cases in which people who have chosen to stay and work with ranchers have been able to restore abused land, create healthier habitat for endangered species, sustain populations of native plants, and more. I would call that neither unnatural nor bad. And I'd take it over a parking lot any day.

CASE STUDY TWELVE

# GROUNDSWELL: COMMUNITY DYNAMICS FROM THE BOTTOM UP

*Todd Graham*

She spread the map on the table before the men; and they leaned in for a closer look. Hay and manure fell from ranchers' coveralls, and the smell of calving filled the room. Melting snow pooled under the boots of government wolf biologists who had just returned from tracking a pack. Placing a finger on the map, she pointed to the high-elevation summer pastures where cattle would soon be located. The biologists agreed that wolves were likely to travel through the area, and the ranchers thought of the safety of their livestock. They looked to her for hope. They needed to unite on this issue and work together through a difficult situation.

Janelle Holden of Keystone Conservation had come to the office of the Madison Valley Ranchlands Group in Ennis, Montana, to explain her idea. It was simple enough: Hire range riders to chase wolves away from ranchers' cattle in the summertime, thereby keeping the cattle alive. But the idea had a huge twist: By keeping the cattle alive, Keystone could save more wolves. By law, ranchers losing stock to predators can request permits to shoot the offending animals on sight. Janelle and her team reasoned that survival rates for both cattle and wolves might improve if a constant human presence were injected into the mix. Keystone was offering to raise money for that presence in the form of a Range Riders Program, provided that the ranchers coordinated and oversaw the endeavor.

Wolves loosed from cages in Yellowstone National Park in the mid-1990s took only a few years to reach the Madison Valley, a 925,000-acre expanse of private and public land lying just off the park's northwest corner. It is easy to see how wolves would prize the Madison, with its herds of big game, open spaces, and clear streams (figure C12.1). The valley is bisected by a world-class fishery in the form of the Madison River, is home to every raptor native to the Northern Rockies, and serves as a migration linkage for big game and large carnivores. When contemplating the Madison, think ecological bounty.

Figure C12.1 The Madison Valley, Montana (photo by Roger Lang).

But the Madison is no stranger to conflict. When they entered the valley, dispersing wolves found cattle and ranchers seeking to protect their livelihoods. A rapidly growing elk herd, which by law belongs to the public, munched privately owned grass. Wealthy newcomers bought ranches and ranchettes, often alienating themselves from the community in the process. Locals accustomed to a certain degree of access to private lands grew frustrated with new landowners seeking to protect their privacy. Retirees moving into the area found nirvana, but long-time locals saw escalating land values and their familiar community and culture fading away. Development occurred, bringing fences and noxious weeds.

It was in this setting of bounty and change that Keystone launched its Range Riders Program, hoping to make a difference. Cowboys were located in various grazing allotments around the valley and were charged with the task of chasing wolves and bears away from livestock. Today, five years later, the program has been successful and used as a model in other areas. Even more important, it has built significant trust among people who historically have not seen eye to eye.

What created the conditions in which ranchers and wolf advocates could agree on a common program? What allowed people to set aside their differences and undertake projects for a larger cause? How did people begin working together, even though they didn't agree on specific issues?

In the Madison Valley, the answer was the formation of the Madison Valley Ranchlands Group (MVRG) in 1996. A group of ranchers gathered under the threat of rapid change in the form of growth and development. They feared for their way of life, knowing revenue from agriculture would not keep pace with land values and operations costs. They felt concern about newcomers bringing values and beliefs they did not understand. In response, they created a community forum through a nonprofit organization where folks of all types could express ideas, voice concerns, and make new friends. They recognized that large divides existed in natural resource management issues and sought means of bringing people together. But how?

MVRG needed a noncontroversial endeavor that united Madison Valley residents regardless of their political beliefs, background, or land ownership status. They found it in the form of noxious weeds, those invasive plant species that affect all land and waterways and whose seeds may be transmitted by livestock and the public's wildlife. They hosted their first noxious-weed fund-raiser in an effort to raise public awareness of the problem and begin an organized control program. The fund-raiser, known locally as the "weed party," has become an annual event in the valley and has raised over $800,000 in less than ten years for various weed-related projects. To date, those funds have been used to hire a weed coordinator, bankroll research programs, purchase weed control equipment that can be borrowed by landowners, and make outright payments to landowners for controlling their weeds.

The program was well funded and celebrated valley wide and built the kind of trust that allowed new management paradigms to be considered. The best of those was the implementation of programs across property boundaries, which in turn allowed the valley's resources to be considered as a whole. A new awareness was generated, causing someone to ask, "Can we treat the entire valley as if it were a single ranch managed for the benefit of livestock, wildlife, and people?"

Fertile lowland soils near the valley's only town of Ennis provide winter pasture for stock owned by longtime valley ranchers. Come summer, those folks haul cows and calves upriver to high-elevation Forest Service grazing lands. Missing from the equation is badly needed spring and fall pasture as cattle transition between the home places on the valley floor to the high country. Traditional ranchers, whose livelihood depends on livestock, need pasture in the mid-valley elevations to make their year-round grazing operations complete.

Those mid-elevation rangelands were purchased in the recent past by individuals who made their money elsewhere. Generally speaking, such "amenity buyers" bought the valley's large ranches for open space and associated wildlife instead of livestock production. Many of those folks come to the valley to recreate, rest, and view the valley's wildlife. Most don't own cows.

An obvious market fit became apparent whereby traditional ranchers seeking grass in spring and fall could graze pastures owned by amenity buyers seeking income from grazing. If a deal could be brokered between the two interests, both might win.

Initially, contrasting values prevented progress. Ranchers needed grass, while amenity buyers wanted healthy wildlife habitat. The challenge became finding a means of using livestock as a tool for improving that habitat. Scientific support from Montana State University and USDA's Natural Resources Conservation Service was enlisted, and the Upper Madison Wildlife/Livestock Partnership was born. Through that program, scientists help create grazing plans and range monitoring systems, while traditional ranchers provide their cattle as the tool for improving wildlife habitat on amenity buyers' ranches (figure C12.2). The result has been that traditional ranchers found badly needed spring and fall pasture, and amenity buyers are secure in the knowledge that they practice science-supported grazing for the benefit of wildlife.

This program has plenty of room to grow by adding more landowners to the mix. Ideally, cattle could be drifted from ranch to ranch on their way to the high-elevation grazing lands rather than being trucked. That would cut traditional ranchers' costs and help them stay in business. In the meantime, the program has contributed to valley wide dialogue and built trust. New doors have opened.

The Madison's allure of wildlife, traditional lifestyles, and scenery has not gone unnoticed. Those whose property values escalated in other parts

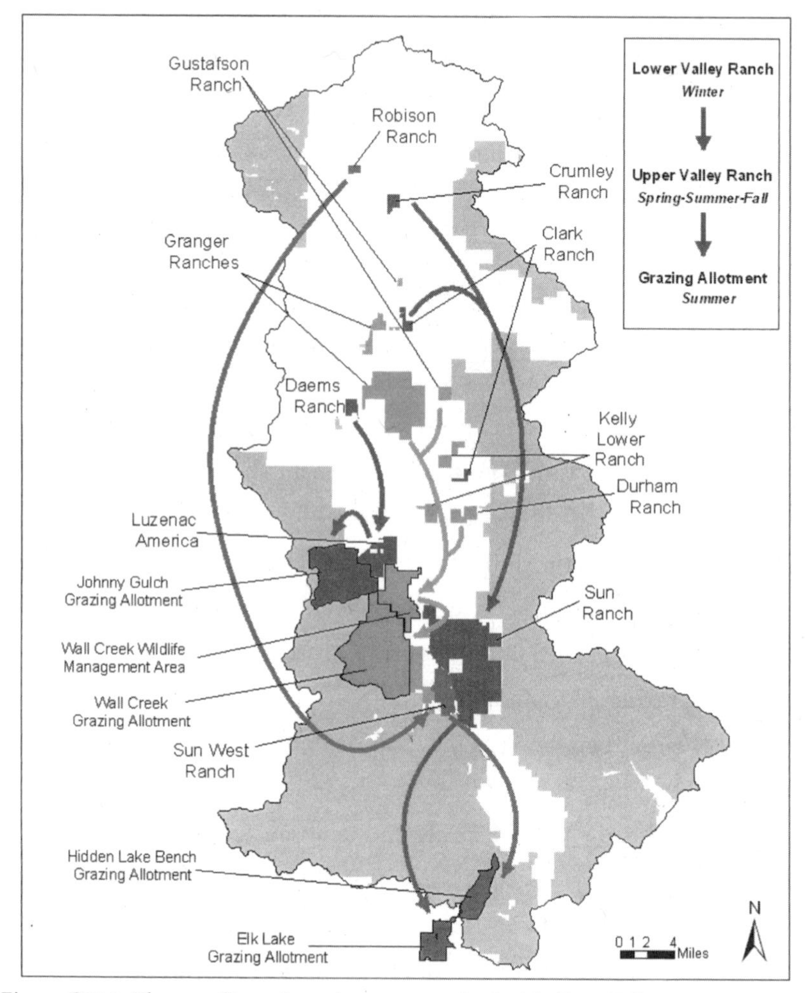


Figure C12.2 The coordinated grazing program in the Madison Valley, Montana (map created by Abigail Brewer, the Madison Valley Ranchlands Group, the Wildlife Conservation Society, and the Craighead Environmental Research Institute).

of the country found comparatively lower real estate values in the Madison and flocked there like snow geese in spring. They came to buy a piece of the idyllic Madison and enjoy its western hospitality. Such growth has filled the pocketbooks of real estate agents, developers, contractors, and merchants alike, while new subdivisions and twenty-acre ranchettes have spread like chicken pox across the landscape.

Realtors marketed the Madison's scenic beauty, openness, and wildlife bounty. The public bought it. Who can blame either side? But open space remains threatened. Houses now stand where elk once grazed. Pasture for cattle is being lost, as are the ancient paths of wildlife migration.

The community needed strategies for managing growth in a way that allowed people a home but also maintained the openness that makes the valley so special. MVRG launched a Growth Solutions process in the form of a community dialogue on human settlement of the landscape. Well-attended forums have been held in recent years, with discussions addressing growth, where growth is occurring, where it should occur, and where conflicts with wildlife and water quality may surface. MVRG was able to capitalize on the trust built by past endeavors to bring people together and discuss the valley's future. While agreement was not always reached, awareness of valley issues was raised. The process helped lead to the passage of a county growth policy, new subdivision regulations, and a community-supported growth action plan. Passage of those documents may help deal with growth issues, but the straightest path forward lies in the new relationships that have been built among people and the new possibilities that those relationships allow.

In the not so old West, newcomers were often welcomed to the neighborhood with baked goods and a Saturday afternoon visit. Such meet-and-greets were how locals indoctrinated new arrivals into the culture. Friends were made, and values were shared. Such practices were easy and fun when only a few new folks arrived each year. But today, as the western states are referred to as America's "third coast" and newcomers arrive in large numbers, locals find that they don't have enough time to bake the cinnamon rolls, let alone go meet new arrivals. A fresh means of assimilating new westerners is needed.

The community in the Madison Valley may have found one good way to welcome newcomers, in the form of Living with Wildlife workshops. Many new arrivals buy their piece of Madison utopia without a basic understanding of the habits of their ungulate and carnivorous neighbors. The workshops are intended to help homeowners in subdivisions understand how to live with four prominent valley species: elk, pronghorn, wolves, and grizzly bears. Participants are taught the principles of wildlife-friendly fencing, why it is important to chase wolves away from your house, why you should put your bird feeders away when grizzlies emerge from hibernation, and other important facets of sharing the landscape with wildlife. At day's end, they enjoy burgers and beer with the locals, including ranchers; MVRG members; Forest Service employees; Montana Fish, Wildlife & Parks personnel; and conservation groups such as the Greater Yellowstone Coalition, Keystone Conservation, and Wildlife Conservation Society. New friends are made. Newcomers get a taste of what it means to be local and share local values. They hunger for more. The community must ensure that those folks continue to learn, build trust, and keep the lines of communication open. By doing so, new possibilities will emerge.

Don't think it's all sunshine and frolicking ungulates on the Madison, though. Don't think everyone gets along. Some people attending valley discussion forums really don't like each other. Further, it is difficult to envision a day when ranchers are fully accepting of wolves, when environmentalists are fully accepting of ranchers, when real estate developers don't pursue that quick buck, and when public hunters are fully satisfied with access to private lands.

But rather than wait on that perfect situation, envision instead a day when people of different interests communicate in a way that fosters resilience and the sense of a cohesive community. Work toward a form of elastic trust. We may not get along on some issues, but we share enough interests to keep talking after disagreeing.

On the Madison, such resilience has allowed the Forest Service to see across the boundary and care for the well-being of private lands, convinced ranchers to care for the well-being of Forest Service lands, inspired Keystone to create the Range Riders Program for the well-being of cows and wolves, fostered communication among conservation groups and real estate developers, and united all those diverse groups in the battle against noxious weeds.

Such efforts provide for the improvement of land health and wildlife habitat and give people the opportunity to make a better living. Folks on the Madison are learning ways to take their base and make it more healthy, vibrant, and valuable. Collaboration and dialogue have allowed for increased awareness of possibilities and a level of resilience that allows future progress.

The Madison Valley is an ecological jewel. Its intactness, openness, and wholeness represent the golden goose. As long as the goose remains intact, she'll keep laying golden eggs and the valley will remain a place where livings can be made and critters can make a home. Though much work is yet to be done on the Madison and threats to progress are evident, enough doors have been opened to form a new question: What can we truly achieve if we work together?

CHAPTER TWELVE

# WHERE WILL THE MOOSE LIVE?

*Bob Budd*

There have been many defining moments in my life, and almost none of them seemed to be important at the time. A great single-malt whiskey takes the right combination of peat smoke, chemical reaction, and time, lots of time. Really important things in life are usually the same way.

My grandparents lived in Wyoming, in a small house on North Piney Creek, on an ancient gravel bar surrounded by floodplains that became hay meadows eighty years before I was born. The narrow lane to their house followed a ribbon of high ground between spring creeks and swamps, and ended on an island of cobble with a house on top. It could take hours to dig a hole there.

The house was pretty typical for its time—small, practical, and inexpensive—built when a major generational shift overcame small ranches and farms. The exterior was ringed by long lines of siding that surrounded the place like stripes on a skunk. Inside, the floors were all a single color of linoleum, and wallpaper covered thin layers of hard, pressed cardboard in most of the rooms. A large propane stove stood in the center of the house.

There were five rooms: a living room, kitchen, two bedrooms, and a bathroom. The warmest place in the house was the bathroom, with an Arvin electric heater running full blast. You entered the house from the south corner of the kitchen, unless you had never been there. In that case you came in from the west, through the front door. One night late, there came a scratching at the front door, and Grandma answered it with a shotgun in her hand. The caller was a pack rat hanging on the screen door, and she blew it to smithereens from about three feet away. She also blew the door to kingdom come.

All she said was "That son of a bitch," like she had been waiting to kill that rat all her life. I never came into the house through that door again.

Everything happened in the kitchen anyway. It was a huge room, lined full length on the east by windows, on the north by a counter that was office for the ranch and home for jars of candy—black licorice, peppermint, and something called horehound that all kids despised. The oven and sink were in the northwest corner; on the west wall were a refrigerator and freezer. The refrigerator held Hires Root Beer, and the freezer stored Popsicles. The house smelled like the kitchen, and the kitchen smelled like meat and potatoes.

For all of the sensory stimuli in that kitchen, the row of windows on the east wall was magical. The roofline was low enough that some winters only a sliver of light would sneak between the roof and the snow on the ground. The sill was about a foot high. If you looked evenly over the bottom sill, you could see under trucks parked outside. In the winter, a smallish boy could look right up at the belly of a full-grown moose, identify the sex, and count the teats. More than often, the moose would look down, poke a broad nose at the top of the window, and steam it from the outside.

I was that smallish boy, and I was absolutely mesmerized by moose.

The moose is the largest member of the deer family, a gigantic, nearsighted, curious beast that appears ungainly and stupid. They were thick as mosquitoes on North Piney, and some winter mornings, I had to wait at the door while Grandpa shooed them away from the pickup. Depending on how much hay was left in the bed of the truck, there might be six or eight moose standing around, their breath combined into a great fog in the subzero air. Grandpa swept them away with a broom, as if they were gigantic metal shavings on the floor of the shop.

I asked everyone about moose. Were they really dumb? How come they were so tame? Should we name them? How many do you think live here? What do they eat in the summer? I harbored the notion that I could climb out the window of my bedroom and ride one when they lay next to the house in the afternoons.

There were answers on the creek. If moose were dumb, they wouldn't be in the yard eating hay. They aren't tame—they just tolerate us. You don't give wild animals names—you give pets and people names. Moose eat moss and willows in the summer. You climb out that window, and it will be the last ride of your life. Watch the cows with calves—they'll kill you quicker than lightning. If they chase you, get in the middle of the biggest willow on the creek and make yourself into a ball—they can't tear the whole willow apart. Don't try to outrun them—they may look awkward, but they're really fast. Over time, I found that almost every single thing I was told about moose was true.

Moose are truly elegant and graceful creatures, both agile and fast when need arises, but generally quite placid, almost tame. They are not a species of herds and mobs but tend to aggregate in smaller family groups. Moose have a fairly steady reproductive rate and have twins when times are good, but it isn't unusual to find unbred cows, or cows with calves from two different years. They don't have a flight response like many large grazers, which makes them vulnerable to hunting. Moose are the ultimate of cool, the kings and queens of the bottomlands, more predator than prey to attackers. They may not be particularly handsome to some species, but they are quite attractive to one another.

Few other large animals are so completely defined by their habitat. Moose live in the narrow ribbons of rivers and streams that divide the sagebrush world of the West. While they may venture into timber and upland, they flourish in places where water is plentiful and out of control. They eat things that grow under the water, and they stand against tall willows to eat the tender growth at the tips of the branches. In the West, where settlement followed water and productive land, you can map moose habitat by cursive biological greenlines or square white lines of land ownership. In the end you have a species that is not particularly prolific, is vulnerable to humans, and is limited by a habitat that is primarily privately owned. Some would argue that such a species is doomed.

And yet, on North Piney, moose remained while other species declined. Mule deer were rare enough in the 1930s to merit excitement. Elk moved in and out of the valley with the seasons but stayed mainly in the higher elevations. Bighorn sheep were present above timberline but not along the creek. Of all the species on North Piney, moose had the most to lose, yet they persisted in the face of every natural and human variation. The anomaly was never a quandary. There wasn't so much intellectual need to explain everything at the time, and, of course, moose were accepted, expected, and loved.

I got my first inkling of why moose persist in an explosive confrontation in the early 1970s. My grandparents were in a financial crisis. The good economy of the 1940s and 1950s was over, and little ranches were falling apart. Both kids had left the place, and good help was nearly impossible to find. North Piney was experiencing only the third shift in generations in eighty years, but it would be much more painful than those before. A dead economy for ranchland and a vibrant market for everything you needed to keep a ranch going had people trapped. Suddenly, everyone focused on how to take a lot more out of the little bit they had. Calve earlier. Fertilize more. Sell calves instead of yearlings. Raise more hay. Kill sagebrush. Cut back. Buy land. Crossbreed. Sell land.

I could no longer look up at the belly of a moose from my gray-and-red, chrome-and-vinyl chair. Snow hid in the shadows of cottonwoods and willows, but the creek ran cloudy, water stood in the meadows, and bluebirds darted on the wind. In the kitchen, Grandpa sat and listened to a government man tell him how to kill the willows along the creek and create more hayfields. The government would help. For every dollar Grandpa spent, they would kick in three.

Suddenly, my normally small, jocular grandfather burst from his chair and set upon the government man. Grandpa snarled and growled and jabbed at the man with his finger, then pushed and shoved him to the door and told him to get out and stay out. He was chased to his truck with the same broom Grandpa used to shoo away moose. In all my life, I had never seen Grandpa raise a hand to another person, ever. He was *mad*.

Grandpa slammed the door and walked over to the stove. He poured a cup of coffee and paced across the floor, an anxious, desperate dance that went one way and then another and finally ended when he sank into a chair at the other end of the table. I followed his movement with my eyes, then looked away when he looked at me and studied the dark liquid in my cup. It isn't easy to see someone you admire go ballistic, and I had no idea what to expect next. Finally seated, he rubbed his eyes and his head, then let out a breath that whistled and wheezed. He crossed his legs and uncrossed them. He turned in his seat and looked out the windows. He tipped his head back and held his neck in his palms, and then he sat up and looked me square in the eye and shrugged.

"Where will the moose live?" he asked.

Where, indeed, will the moose live? My family helped to settle this valley, more than a century ago. They helped to make the watershed a home for many. They carved a life from swamp and grass and sky and snow up to a tall man's ass. They cussed snow and they cussed drought and they cussed each other, and then they loved snow and each other. They learned to love wind and cold, and they sank their own roots in the rocky ground and hung on to the place they knew and loved. And if they hadn't loved moose, there would be no moose.

The only reason moose are still abundant on North Piney is that the people who live there want them there.

Aldo Leopold once said, "There is no conceivable way the general public can legislate crabapples, or grape tangles, or plum thickets to grow up on these barren fencerows, roadsides and slopes, nor will the resolutions or prayers of the city change the depth of next winter's snow nor cause cornshocks to be left in the fields to feed the birds. All the non-farming public can do is to provide information and build incentives on which farmers may act."[1]

Gene Decker taught wildlife biology at Colorado State University for decades, and he tells anyone who will listen that the most important skill a wildlife biologist can learn is to "hunker." He demonstrates, squatting on his haunches and poking at the ground with a stick. He is adept at listening, a master of *feeling* the words and emotions of those with whom he hunkers. He has hunkered all over the world. He hunkers sometimes when he could sit in a chair. Gene has large ears, an open mind, and a heart the size of a moose.

Ultimately, conservation is vested in the heart. No amount of science, regulation, or wishing can make moose pop out of the bog. Conservation is sensual, and personal. Without the will to conserve, action lacks passion. Brain may connect with muscle, but blood still has to flow through the heart.

Conservation is not a battle. It is not the end game. There is no winner at the "end" of conservation. Conservation is *always* a means to an end. The result of conservation is three dimensional, abstract, and often unknown. The glory lies only in the pursuit, never in the capture. When there are losers in conservation, they are usually everyone involved.

Conservation happens when things make sense. Things make sense when they do no harm or potentially improve on the existing situation, when they offer some source of satisfaction, without a sense of worry. Regulation, always based on "science," too often converts sensibility to threat. When that happens, the desire to conserve vanishes like cold breath in a hot kitchen. Statistics and science are important but not when they create nagging worries in the minds of those on the land. Too many have been offered that chance to "kill the willows."

Conservation is driven by personal investment in the past and the future, and nothing inspires like the combination of heritage and succession. For most species, the combination of "here and now" gives way to the future only for brief moments of courtship, rut, and spawn. Humans have the capacity to move mentally in three dimensions. People on the land constantly weigh a sense of duty to the past with the needs of the present and the hope of the future.

The frailty inherent in that ability comes when all dimensions play into a single moment. The bond to place is a hard knot, less often unwound than cut. When you leave a place you love, it is hard to go back. Decisions that change the place you love are no less painful. They are agonizing and sometimes final. It matters little whether you are rancher, ranger, or game warden. Those who are new on the landscape must try to comprehend generations of duty, a season of drought, and a legacy left uncertain, sometimes all at once.

The temptation has become a tendency to resort to science, rely on statistics, and assume we are right. In doing that, we bypass the human element, the heart and soul that ultimately drive the survival of creatures like moose. If we really want to achieve landscape-scale conservation, we must first lose all vestiges of bias and disdain we learn through examples of failure. We have to learn to hunker. We have to grow bigger ears. We have to comprehend the commitment of local people to the landscapes we wish to maintain.

The greatest challenge in wildlife conservation lies not in our ability to radio-collar and track moose on North Piney Creek. It is not our inability to accurately map moose habitat within meters of reality. We can model populations of moose and guess at their future output. We are especially good at understanding the things moose need to survive, but only recently has natural resource science begun to include the human element in the equation. For all the vagaries and nuances inherent in ecology and biology, understanding the attitudes of people toward species is even more complex.

Most people who work in conservation today didn't grow up in a kitchen on the creek where they work. Nearly all bring a combination of passion and good science to the table, but the focus is generally on science. For a period of time, we believed that was enough. We made decisions and recommendations based on today and the future, without appreciation and understanding of the past. Thus, we lost or ignored the knowledge and passion of those who live in the willows.

At the same time, people on the land were dealing with myriad distractions—markets, weather, animal health, agency regulations, family, new agency personnel, new species, new ideas, new neighbors. Between 1980 and 1990, even the most progressive businesses struggled to deal with conversion from the IBM Selectric typewriter to the laptop computer, from Volkswriter to WordPerfect to Office. If not so spectacular as the transition from horse to spaceship seen by people still living on the creek, the explosion in communication and analytical process was unequaled and much, much faster. The kitchen on North Piney still had a perfectly good manual typewriter on the counter, blank paper, and lots of pens in a jar, and all of a sudden, they were irrelevant.

Wildlife biologists were caught in the same vortex. People who built a career in the field were hit with paperwork and computer models and public hearings. The new idea was that science would answer every question. Congress wanted to hear about science. The Supreme Court was interested in science. *Everyone* was absorbed with science, as if suddenly the interest would make the natural world conform to the rigors of physics and chemistry. Surely, the combination of computers and science would prevail. Some used science in a manner that was openly hostile, arrogant, and demeaning, but mostly, it was meant to be helpful.

Up on North Piney, the moose missed the lecture.

So did the people who loved the moose. The lack of history by one partner collided with information overload in the other. It was as if one dancer heard a raucous jig while the other tried to count a waltz in her head. Only passion held them together, gave them the will to comprehend and accept different rhythms and dance. But in many cases, the dance just ended. The music stopped. The community hall was empty. Passion gets dancers on the floor; after that, chemistry becomes much more than science.

The most critical thing we can do to conserve natural resources is to comprehend the past from people like those who love moose on North Piney Creek. All the computers and models and science in the world cannot capture the passion and the understanding in the valley. You feel the land when you are on it every day. You find multiple tints of magenta in sunsets. You learn distinct hues of green on different species of willow. You see infinite shades of white in a drift of snow.

The human aspect of conservation is the most essential, and perhaps most ignored, of all disciplines. Only in the past five to ten years has it even become accepted as a field of study. The collective knowledge of the people on the North Piney creeks cannot be divined by modeling, wishing, or study. There is only one way to capture the wisdom of people, and that is by listening to them.

The best listening comes in warm kitchens on creeks where moose steam the windows with their breath. The human mind becomes three dimensional at a table with a heritage. Most important, you begin to realize that there are real reasons for the things people do, for what they believe and what they feel. Some of those reasons may be fragile, even unfounded, but they are reasons, just the same, and they are important. Ultimately, few species can survive against the will of the people who live with them every day. Bobcats in the willows are welcomed; bobcats in the barn are not. Much of conservation lies in a delicate balance of mutual risk and faith.

The people you meet when your knees are under their table are an eclectic mix and instantaneously shatter all images of who does conservation on the ground. There are octogenarians who never left the county, gazillionaires who have been there for only a few years, young couples trying like hell to resist the money of the city and raise their children in the willows. Most are very well educated, and some are downright brilliant. When they find commonality in doing right by the land, you feel honored to sip their coffee.

Not long ago I asked four very different landowners on North Piney how the moose were doing. All told me first about a huge cow with twins, a moose that will likely be the comparison for all females to come. David had seen the big cow with twins the previous morning at the Lewis Place. Chad had run a dry cow and two little bulls from the Noble Place on Thursday. Cotton said there was a bull with antlers like a caribou down at the Beck Place, and he had a smaller cow with twins with him. "That bull was up in my field this morning," Bill said, "but he had a dry cow and a pair with him." David had seen the dry cow between the Muir Place and the Noble Place the night before. She was coming in heat, he figured, and the caribou bull must have chased the younger bulls off.

Between the four of them, those landowners had read about everything available on moose ecology, and they followed the animals with an intimacy generally saved for children. They knew most of them by sight, how many should be around, and where. I asked all four how many bulls were in the valley, about fifty square miles of country, and their estimates were uncannily close. When I wondered if that was enough bull moose, all said they would like to see more.

While questions are often addressed with incredible insight, suggestions about habitat improvement are commonly met with a quiet stare, no commitment, and questions. It is a common process, one that takes time to evaluate credibility, risks, and the reliability of making changes on the land. When those things make sense, the response is swift and anxious. Reverence for the past and commitment to the future become the present. The waltz becomes a jig.

One of the great frustrations in working in that setting lies in getting the step to the music right, at the right time. More often than not, landowners will take a very deliberate approach to conservation proposals at the outset, especially if they imply changes that may be detrimental. When those landowners decide to move forward, they suddenly hear an anxious backbeat and wonder why their desire to dance is being ignored by others.

One family decided to place a conservation easement on their ranch after thinking about it for more than two years. In the space of that time, one or another of them might call about every six months. They searched the Web; read books; and talked with their accountant, every member of the family, and anyone else who might be able to shed light on every conceivable option. Once they decided it was the right thing to do, they called weekly to see why it hadn't happened. They were worried, and they were anxious. Above all, they were excited about what they could do for a place they loved, their own home.

On the other end of the spectrum, conservation partners heard rock and roll from the beginning. The science said yes. The habitat was fantastic, thanks to the care of the family on the land, but its future was tenuous, vulnerable to being converted from ranch country and habitat to houses. The loss of families on the land was troubling. Agencies were ready to move when the sun came up in the morning. They interpreted hesitance as unwillingness and assumed they had lost the opportunity for conservation. By the time the family was ready to go, agency personnel were looking for funding, trying to get appraisals and biological assessments, seemingly moving to the beat of a funeral dirge.

Every time you dance with someone new, it takes time to adjust to another set of muscles. Some hear the drum. Some hear the bass. Some hear a sound you cannot comprehend. You dance again for many reasons, but rarely in complete harmony. Mostly, you dance because you share a passion. Sometimes the passion is for each other. It may be the music, step, or rhythm. There are also times when you dance together because you both love something else.

We have made incredible advances in science and technology in the past few decades. We can build maps from outer space or from palmtop computers on the back of a horse. We can track moose from deoxyribonucleic acid, ATVs, satellites, or moose poop.

Were I a moose on North Piney, I believe I would place my trust in Bill, Chad, David, and Cotton. I would stare into their kitchen windows and try to focus on smallish boys. I would hope like hell everyone at the table had the same passion for each other that they had for me.

**Note**

1. A. Leopold, "Game Cropping in Southern Wisconsin," in *Aldo Leopold: The Man and His Legacy*, ed. T. Tanner (Washington, DC: Soil and Water Conservation Society, 1995), p. 42.

CONCLUSION

# AN UNPRECEDENTED FUTURE

*Courtney White*

Not long ago, Craig Allen, an old friend and colleague, stopped by the office to catch up. He is a forest ecologist stationed at Bandelier National Monument, in northern New Mexico, and his career is representative of the transition conservation has undergone in the past decades, as well as its likely future trajectory.

When I first met Craig, nearly twenty years ago, his focus was on ecological change at landscape scales—in that case, the Jemez Mountains, in which Bandelier is nestled. His approach was a systems one—he studied the interlocking variables of ecological function, historical use, and plant and animal community dynamics in order to understand more clearly the condition of the forest. And what he discovered was worrisome. Specifically, he worried about forest "thickening" due to decades of fire suppression, overgrazing, and other human activities.

In 1998, Craig summarized his concern in an article titled "Where Have All the Grasslands Gone?"[1] His research revealed that open, grassy areas in the Jemez Mountains were shrinking, due to tree encroachment, at the alarming rate of 1 percent per year. What was missing was fire.

"Most forests, woodlands, and grasslands in northern New Mexico evolved with frequent, low-intensity fires," he wrote. "The removal of the natural process of fire by human suppression has disrupted these ecosystems in many ways. . . . [These areas] need to be restored to more open conditions to protect both ecological values and human communities."[2]

In the next phase of his career, Craig "walked the talk" of forest restoration by implementing innovative experiments at Bandelier, becoming an enthusiastic advocate of adaptive management in the process.

One summer, Craig gave me a tour of one of his projects. It incorporated two microwatersheds, situated back to back. The first one was left alone as the control, while in the other one crews with chainsaws delimbed piñon and juniper trees and spread the branches and needles over eroded patches of bare soil. The goal of this restoration work was to slow erosion, and the outcome was impressive. Less than two years later, grass had begun to grow in the spaces between the branches (because the branches checked the flow of water, allowing infiltration), slowing soil movement. The control site, by contrast, continued to erode.

As a consequence of this and other fieldwork, Craig joined a chorus of forest ecologists advocating proactive policies and practices aimed at returning ecosystems to health in the Southwest, principally by restoring natural fire cycles.

In other words, Craig Allen's work is typical of the new approaches to conservation detailed in this book: Protection has given way to restoration; the isolationism of parks and wilderness areas has given way to integrated approaches to whole landscapes; solitary research perspectives have given way to multidisciplinary strategies; a simple array of land conservation tools has given way to a complex toolbox.

But Craig's story doesn't stop here—and neither does ours.

Today, Craig's focus is on the dangers posed to forests by global warming. He thinks the dangers have the potential to be catastrophic not only for trees but also for the animal communities that depend on them—including us. "The possibility exists," he told me that day in my office, "that a 5 degree Celsius warming of the planet could wipe out entire plant communities, including the forests."

He also thinks time is short, which is why he has joined other scientists who advocate a reduction in $CO_2$ emissions, pronto.

But it was something that Craig said at the end of our meeting that brought home the question of conservation's future for me. Not long ago, he said, he had an opportunity to speak to a gathering of federal land managers about the climate crisis. They were looking for options and advice on how to meet that challenge.

"What they told me," Craig said, "was that nothing in their education or experience had prepared them for what was coming down the road in terms of climate. Their training was for a stable climate, they said, not one that was changing. They literally had no idea what to do. They were facing an unprecedented future for which they were not prepared."

Which brings me back to this book: Conservation is certainly in transition, as the authors gathered here ably explain—but in transition to *what*?

* * * * *

For most of his career, Craig Allen focused on a traditional goal of the conservation movement: fighting scarcity. Unhealthy forests, disappearing meadows, eroding topsoils, too few "cool" natural fires, too many "hot" catastrophic fires, and not enough grass are all indicators of scarcity at work—the scarcity of properly functioning ecosystems. His restoration work aimed at reversing such declines, at replacing scarcity with health and abundance.

Today, however, Craig is working beyond scarcity. He is confronting the specter of *loss*. His work now is not simply to reverse scarcity, but to prevent loss. In that, Craig's effort is emblematic of an important shift in conservation. The main goal of the movement over the decades has been to *conserve* those parts of nature threatened with scarcity, whether a place, a species, an experience, or an ecological service that humans value for aesthetic, recreational, ecological, or economic purposes.

A classic case is the creation of the national forest reserve system in the United States at the end of the nineteenth century. Appalled (and worried) by the industry-fueled razing of the great white pine forests of the upper Midwest after the Civil War, Congress created what eventually became the national forest system and gave it the mission of conserving valuable timber for future national consumption. The specter of a timber famine, in other words, provided motivation for conservation action.

So it went with national parks, national wildlife refuges, wilderness areas, wild and scenic rivers, endangered species, wetlands, water, clean air, open space, farmland, biodiversity, and so on. In nearly every case, scarcity of one sort or another played an important role in provoking a conservation response. And as the nature of the scarcity evolved over the decades, so did the methodology of conservation.

But today we stand on the precipice of something new: the possibility of the irreversible loss of big things we value in nature. Cut-over forests grow back, especially if the right conservation strategies are employed, but climate-induced extinctions are forever.

On the face of it, this fact presents us with a dispiriting situation—but not a hopeless one. Wendell Berry summarized it well in his essay "The Whole Horse": "My hope is most seriously challenged by the fact of decline, of loss. The things that I have tried to defend are less numerous and worse off now than when I started, but in this I am only like other conservationists. All of us have been fighting a battle that on average we are losing, and I doubt that there is any use in reviewing the statistical proofs. The point—the only interesting point—is that we have not quit. We have not quit because we are not hopeless."[3]

The authors presented in this book have not quit. On the contrary, they have given us hopeful examples of action. In the transition from the goal of protecting scarce resources to that of stemming outright loss, conservation has developed and implemented a diverse suite of innovative, collaborative, and effective strategies, many of which are detailed in this book. And in their success, they point the way to further innovation and implementation.

The crash of fish stocks off the coasts of New Zealand, as described by Dewees, is a good illustration. By the mid-1980s, private and public conservation strategies had proven themselves inadequate to prevent the impending collapse of native fisheries. Confronted with the prospect of huge economic and ecological loss, an entirely new strategy was devised—one that required new partnerships, new incentives, and new management strategies. And it worked.

Jones and Kochel tell similar success stories about wildlife, though in the case of the Kinzua Deer Cooperative in Pennsylvania, the crisis was one of overabundance, not scarcity. In both cases, natural systems had become disordered to the point of near crisis, requiring deeper conservation responses than had been used in the past.

Another good illustration of the transition from scarcity to loss is the response to the conversion of open space, including working farm and ranch landscapes, to subdivisions and other forms of land fragmentation. Fifty years ago, this concern barely registered as a conservation priority, but today it has become the top priority to many of us in the United States, as well as other parts of the world.

The authors detail effective and inspirational responses to this crisis. Sherrod explains the important role that land trusts and conservation easements play in protecting ranchlands from development. Adams argues that parks do a similar service in a different type of landscape by preventing land conversion. Parks can also do more, he writes, including carrying out landscape-scale conservation work in collaboration with neighbors and diverse partners.

The chapters by Hilty and Groves, Lucero, and Propst and colleagues demonstrate other successful and innovative approaches to stemming the threat of land fragmentation, including bottom-up planning that engages multiple stakeholders, planning at scales that cross administrative boundaries, engaging diverse public institutions and political leaders, providing tax and market incentives that encourage land preservation, and developing community stewardship organizations that work at the local level.

Ginn provides a fascinating look at how a new strategy, which he calls doing conservation "at the speed of business," can effectively protect forests in New England from development. Similarly, Kelly takes us on a tour of the opportunities and challenges that arose when a nongovernmental organization not only moved speedily to protect critical forestland in northern California but subsequently embarked on a novel experiment to manage the land for economic as well as ecological goals. Such goals are familiar to Andrew, who explains in his case study how some farmers have been working at the nexus of ecology and (no-till) agriculture for a while now.

This trend of conservation groups managing land for sustainable use—what I call "conservation with a business plan"—will grow significantly in the twenty-first century. I think of it this way: If ranchers and timber producers are becoming more like conservationists, shouldn't conservationists become more like ranchers and timber producers? I think they should. In fact, combining the approaches of those two groups likely holds great promise for the future of conservation. Keeping land open and unfragmented while also maintaining its productive capacity is good for both people and nature.

Of course, conservation ranching, and healthy business practices, is nothing new to the James family, who have protected a significant parcel of open space north of fast-growing Durango by *doing* conservation for many years. But their success rests not just on best management practices or a sound business plan. It rests on solid relationships and a clear sense of values, which typically (and appropriately) resist quantification. Some things exist beyond numbers. The Jameses put it best: "When human dignity or worth is diminished, the land and the economy will eventually fall into misuse."

Strong relationships are the foundation to another key strategy: collaboration. As Snyder, Dagget, and Graham explain, collaboration between diverse, and sometimes historically antagonistic, partners can lead to long-term solutions to natural resource challenges. Whether it is at the scale of a community, a large valley, or peace-making between ranchers and environmentalists, the keys to effective conservation today are respect, communication, and creativity. The new conservation "toolbox" can't be effective if people are not willing to work together to give the tools a try.

Other forms of relationships are critical, too—such as the political and social divides that have long separated urban and rural populations. Knight explores the efforts to bridge those divides, especially around the issue of local food. In the process, he expands on an experience that the James family discovered early and many us of are just discovering now: Food can be the foundation to long-lasting conservation. Because eating is an agricultural act, as Wendell Berry reminds us, and since agriculture involves conservation at so many levels, food becomes a critical point of contact for a dialogue between urban and rural people. And once dialogue starts, relationships follow.

The loss that is being redressed in all of this, as Forbes explains, is the loss of a culture of care and attention to the land: In the early days of conservation, he notes, we attempted to keep safe what we valued by removing people and certain species that were perceived as threatening. For example, the famous conservationist John Muir, Forbes recounts, rocked his canoe so his Tlingit guide couldn't shoot a deer on shore. That has changed. Today the challenge is to find ways to reconnect people to the land, principally through work. We must also honor cultural diversity in all its forms. "Resilient relationships deserve our respect," Forbes writes. We need to learn from others and not just pat ourselves on the back for our ingenuity. The goal of conservation, he concludes, is not to work at bigger scales, or at faster speeds, but to "engage the hearts, minds, and everyday choices of diverse people."

In her case study, Crona identifies a key aspect of relationships: leadership. Since surprise and change are inevitable in nature, continuous learning is a prerequisite for sustainable management, she observes. Those individuals in a village or community who not only have high social skills but who are willing to keep absorbing new knowledge often become the leaders. Conversely, the loss of such a willingness to learn can lead a village or community into trouble. But it is not just about individuals—the entire community must learn together so that it can manage a natural resource collaboratively. She calls this strategy "adaptive comanagement." It is a strategy that we should begin to consider here in the United States, perhaps.

Leadership and learning are also the focus of the chapters by Cheng and Kemmis, who examine those issues through the lens of public lands. In the process they explore a different sort of loss—the increasing inability of public agencies to deliver public goods and services. In the case of the U.S. Forest Service, the subject of Cheng's inquiry, that inability arises from the loss of employees—staffing levels have dropped 25 percent in eight years. But it extends further than that. Federal agencies are struggling to meet rising and diversifying social demands on public lands because they lack both a capacity and a willingness to enter into *reciprocal* relationships with communities.

One answer to the trouble with public agencies, Kemmis argues, is to give innovation a chance on public lands. "There are no road maps," he writes, referring to the uncharted territory that a deliberate period of experimentation would entail. Creating the road maps would require political bipartisanship—which Kemmis believes to be a prerequisite for conservation in the twenty-first century. Politics matter, he reminds us, and policy matters. It can't all be grassroots. To innovate effectively, to chart new paths and create new relationships, the role of national and regional governance structures must be reconsidered.

So must the law, as Ruhl explains in his chapter. The social and economic benefits of ecological services—services just now becoming understood as critical to the future of human life on the planet—have for too long been implicit in the law, he writes. Now they need to be made explicit. Ecosystem services (air, water, wetlands, soil) are nature's capital and are *not* free, as we have supposed under classical economic theory. The economic costs of managing them poorly are rising fast, and legal strategies need to be figured out so they won't be lost. Partly that means market forces, partly that means legal protection—either way, time is short. Of course, commodification of ecological services carries risks, Ruhl notes, but conducting business as usual will only make matters worse.

Thus we reach an important commonality of all the pieces in this book: Business as usual can't continue.

* * * * *

In all the models of future scenarios that I have seen—for climate, energy, population, food, water, and growth—business as usual (BAU) is the doom-and-gloom alternative. Take global warming: BAU cooks the planet. We must do something else. This applies to conservation, as well. Yesterday's conservation strategies have become today's BAU. When it was clear that the old strategies couldn't handle the transition from scarcity to loss, new approaches to conservation were developed.

But that raises a critical question: Can the new approaches adequately prepare us for the unprecedented future that is relentlessly approaching?

Possibly.

In their introductory and concluding chapters, Meine and Budd, respectively, articulate an important, overarching message: We must learn to live within limits. We must learn to restrain ourselves. We must do so not simply because it is virtuous, as conservationists have argued for decades, but because it is now necessary. The specter of loss, which has moved rapidly from ghostly concern to material reality, now requires us to do something that we have thus far refused: to make hard choices. If we want moose in the woods, as Budd's grandparents did, then we have to learn to "hunker," as he puts it. We can't have it all anymore. We have to *choose* our future, and in choosing we must be prepared for sacrifice.

As conservationists, we understand intellectually what the Cree elder in Meine's chapter means when he says we are all "grassroots managers of this vulnerable ecosystem." But in our daily lives we have rarely been willing to practice what is necessary to accomplish this important vision of stewardship and responsibility. Over the decades, we have made the "easy" choices: parks, refuges, wilderness areas, chemical bans, land trusts, tax incentives. But as the vulnerable natural world continues its transition from scarcity to loss, as we enter an unprecedented future for which we feel unprepared, we are faced with more difficult decisions.

They are difficult because they are economic, not environmental, decisions. As Wendell Berry and others have repeatedly pointed out, our environmental crisis is at heart an economic crisis. A bad economy has bad consequences, as we are rapidly discovering. Addressing the symptoms without addressing the causes of loss, as we have been doing for a long time now, is, to paraphrase Aldo Leopold, like "fixing the pump without fixing the well."

We are at the point at which the well must be repaired. And this will be a hard chore for the conservation movement—because it has traditionally kept the economy separated from the environment in its work.

In "The Whole Horse," Wendell Berry put it this way: "The conservation movement has a conservation program; it has a preservation program; it has a rather sporadic health-protection program; but it has no economic program, and because it has no economic program it has the status of something exterior to daily life, surviving by emergency, like an ambulance service."[4]

He goes on to say that the movement's greatest failure has been its unwillingness to help create an economic alternative to industrialism.

A good economy—argues Berry—will be place based; will work within natural limits; will provide care and respect for nature; will emphasize food, family, and community.

It is not so far-fetched. Already, there are hopeful, early signs of how this new economy might work. I have seen it in action, albeit on small scales, on the ranches and farms around the nation that have chosen to "deindustrialize" by going back to the original solar power: photosynthesis, returning to a land health paradigm (demolished by industrial agriculture) that emphasizes the grassroots—the grass and the roots. I have also seen an interest in sustainable agriculture grow among a smattering of conservation groups—groups that are healing land through restoration work, building relationships, and even producing local food themselves. I call it a "back-to-the-future" economic model—a new agrarianism, but with laptops and cell phones.

While the details of this new economy lie beyond the scope of this book, I think considering its implications takes us a long way toward answering the question, *Transition to what?* I think: to a new economic arrangement. Reversing loss—the next step after stemming it—means we need to work on the pump *and* the well. That means we need a conservation movement that works at the nexus of agriculture and ecology—growing, restoring, sharing, regenerating, and sustaining—while *profiting* from our good work.

Is it possible? I have no idea. No one does—that is why it is an unprecedented future. But BAU requires that we climb into the well and try to get it working again.

At the same time, we need to continue the struggle against loss, as Craig Allen is doing when he works for a reduction in greenhouse gases. It is a struggle that most of the authors in this book are engaged in, as well—keeping up the good fight. Ultimately, the future of conservation means integrating both efforts: preventing loss while exploring a new economic model. They are not mutually exclusive projects.

What does all this mean for the next generation of conservation leaders? I'm not sure anybody truly knows. All we can say with confidence is that the challenges as well as the opportunities are legion. But a new generation has an important advantage: It stands on the shoulders of the one that went before it. And this particular upcoming generation has the advantage of standing on impressive shoulders.

The view, I'm sure, will be inspiring.

**Notes**

1. C. Allen, "Where Have All the Grasslands Gone? Fire and Vegetation Change in Northern New Mexico," *Quivira Coalition Newsletter* 1, no. 4 (May 1998).
2. Ibid., p. 17.
3. W. Berry, "The Whole Horse," in *The Art of the Commonplace: The Agrarian Essays of Wendell Berry*, ed. N. Wirzba (Emeryville, CA: Shoemaker & Hoard, 2002), p. 244.
4. Ibid., p. 246.

# LIST OF CONTRIBUTORS

**Jonathan Adams** is a conservation biologist and writer with The Nature Conservancy. For the past two decades he has worked extensively in Africa, Asia, and North America, focusing on the role of communities in the conservation and management of protected areas. He has also helped ensure that conservation scientists and practitioners have free and open access to the information they need to make effective decisions. He is coauthor of *The Future of the Wild: Radical Conservation for a Crowded World* and *The Myth of Wild Africa: Conservation Without Illusion* and coeditor of *Precious Heritage: The Status of Biodiversity in the United States.*

**James O. Andrew** is president and general manager of Andrew Farms, Inc., a family farm near Jefferson, Iowa. He has been a director and officer of the Iowa and National Corn Growers associations, Iowa Corn Promotion Board, U.S. Grains Council, and Iowa and American Soybean associations during his thirty-four-year farming career. He has used those positions to promote farm policy and conservation issues worldwide. In addition to scoring in the top tier of the federal Conservation Security Program, Andrew was named the 2007 national "Conservationist of the Year" in the Conservation Legacy Award program sponsored by the *Corn and Soybean Digest*, Monsanto, and the American Soybean Association.

**Bob Budd** is the executive director of the Wyoming Wildlife and Natural Resource Trust, the State of Wyoming's wildlife habitat and conservation funding program. He has previously managed the Red Canyon Ranch for The Nature Conservancy and served as executive director of the Wyoming Stock Growers Association. He has degrees in range management (MS), agricultural business (BS), and animal science (BS) from the University of Wyoming. He is past president of the international Society for Range Management and currently serves as president of the Wyoming Chapter of the Wildlife Society.

**Nina Chambers** is director of programs for the Sonoran Institute. She leads a collaborative conservation partnership between the institute and the Bureau of Land Management to identify best practices for community partnership in public land issues. She also helped develop regional ecosystem management approaches in the Sonoran Desert.

**Antony S. Cheng** is an associate professor in the Department of Forest, Rangeland, and Watershed Stewardship at Colorado State University, where he teaches, conducts research, and performs outreach in forest and natural resource policy. His ongoing interest is in collaborative stewardship and community-based conservation. He has been involved in numerous community-based, collaborative stewardship efforts on national forestlands, has provided capacity-building trainings to enhance collaborative practices, and is coauthor of *Forest Conservation Policy*.

**Beatrice I. Crona** is a senior researcher at Stockholm University. Her background is in marine ecology and coastal restoration. Her recent research has focused on management of natural resources in coastal regions, with a special interest in local ecological knowledge and how it is generated and transmitted though social networks.

**Dan Dagget** is an author, public speaker, and consultant on collaborative conflict resolution and restorative land management. His first book, *Beyond the Rangeland Conflict: Toward a West That Works,* was nominated for a Pulitzer Prize. His most recent book is *The Gardeners of Eden: Rediscovering Our Importance to Nature.* He currently lives in Santa Barbara, California, with his wife, Trish Jahnke.

**Adam Davis** is the president of Solano Partners, Inc., a consulting firm focused on environmental investment and conservation finance issues. He is a partner in Ecosystem Investment Partners (EIP), a private equity management firm that acquires and manages high-priority conservation properties across the United States. He is a cofounder and previous editor-in-chief of the *Ecosystem Marketplace*, a global information service on market mechanisms and financial incentives for conservation, and a member of the advisory council for the Aldo Leopold Leadership Program, which provides training for environmental scientists across the United States in public speaking and communications.

**Christopher M. Dewees** is the marine fisheries specialist (emeritus) at the University of California at Davis. He has worked primarily on fisheries management issues since 1972 with special emphasis on innovative approaches to improving the economic and ecological performance of ocean fisheries.

**Peter Forbes** is a photographer, writer, and lifelong student of the relationship between land and people. After eighteen years working in land conservation, he began in 2002 to unfold an ambitious dream of creating a new type of social change organization, one grounded in place but diverse in its relationships, one that seeks to understand and create healthy, whole communities. Today, the Center for Whole Communities has alumni from forty-five states and more than 450 communities and organizations. Forbes's essays on land, people, and culture have appeared in a dozen books, including *Our Land, Ourselves: Readings on People and Place; The Great Remembering: Further Thoughts on Land, Soul and Society;* and *Coming to Land.*

**William J. Ginn** is a businessman turned conservation practitioner who has helped The Nature Conservancy protect over 2.5 million acres of forestland through dozens of innovative deals. He is currently director of conservation markets and investments at The Nature Conservancy. A human ecologist by training, he has served as the executive director of the

Maine Audubon Society, founded and was president of Resource Conservation Services, Inc., has worked as a business consultant, done mergers and acquisitions for major public companies, and, for the last ten years, worked for The Nature Conservancy on forest conservation projects.

**Todd Graham** is the president of Aeroscene Land Logic, a Montana-based firm providing ranch management, grazing planning, and rangeland health monitoring services to landowners and livestock producers. He currently serves as board chair to the Greater Yellowstone Coalition, a 12,000-member conservation organization based in Bozeman, Montana. He is coauthor of a rangeland monitoring and management manual called *Bullseye! Targeting Your Rangeland Health Objectives.*

**Craig R. Groves** is director of conservation methods and tools for the Conservation Strategies Division of The Nature Conservancy. During the writing of his chapter, he was a senior conservation scientist for the Wildlife Conservation Society. During his tenure there, he managed several wildlife and conservation research projects in the Greater Yellowstone region and provided technical support on conservation planning to international projects. He coauthored *Drafting a Conservation Blueprint: A Practitioner's Guide to Planning for Biodiversity.*

**Jodi A. Hilty** is a landscape ecologist with the Wildlife Conservation Society (WCS) with a research focus on understanding thresholds of human impact on biodiversity. In her capacity with the society she oversees a diverse program of conservation and research efforts in North America. She is lead author of a book titled *Corridor Ecology: The Science and Practice of Linking Landscapes for Biodiversity Conservation.*

**Kay and David James** have been ranching in the Animas Valley of Durango, Colorado, since 1961. Their five children and families manage the ranch as a whole to support individually owned enterprises of 100 percent grass-fed cattle, a seasonal dairy and artisan cheese production, pastured pork, organically grown vegetables and fruits, pasture-produced eggs, and a landscape tree business. The products from those businesses are direct-marketed from Durango to the local four-states area.

**Mike Jones** was born and raised on a farm in the United Kingdom and served in Zimbabwe's Department of National Parks and Wildlife Management from 1973 to 1995. He joined the Sand County Foundation as director of the Africa component of the Community Based Conservation Network Program in 2001. The goal of that work is to develop the institutions and organizational capacity necessary for adaptive and sustainable use of natural resources at a landscape level.

**Chris Kelly** is the California program director for the Conservation Fund. Prior to joining the fund, Chris assisted a number of land conservation organizations, including the Save-the-Redwoods League, the Yosemite Fund, and the Elkhorn Slough Foundation, with program and project development in California. For twelve years ending in 2000 Chris worked for The Nature Conservancy of California, where he led the design, development, and implementation of innovative conservation projects throughout California. He holds a BA in philosophy from the University of California at Berkeley and a JD from the University of California at Davis.

**Daniel Kemmis** is a senior fellow and past director of the University of Montana's Center for the Rocky Mountain West. He was formerly mayor of Missoula, Montana, and speaker and minority leader of the Montana House of Representatives. He is the author of three books: *Community and the Politics of Place; The Good City and the Good Life;* and *This Sovereign Land: A New Vision for Governing the West*. He has been the recipient of the Society for Conservation Biology's Distinguished Achievement Award for Social, Economic and Political Work and the Center of the American West's Wallace Stegner Prize for sustained contribution to the cultural identity of the West.

**Richard L. Knight** is a professor of wildlife conservation at Colorado State University. He was selected by the Ecological Society of America for the first cohort of Aldo Leopold Leadership Fellows, which focus on leadership in the scientific community, communicating with the media, and interacting with the business and corporate sectors. Presently, he sits on a number of boards, including those of the Colorado Cattlemen's Agricultural Land Trust and the Quivira Coalition, as well as the Science Board of the Malpai Borderlands Group and The Nature Conservancy's Colorado Council. In 2007, Colorado State University awarded him the Board of Governor's Excellence in Teaching Award.

**Jeffrey Kochel** is the northern operations manager for Forest Investment Associates and is responsible for timberland management on properties in Pennsylvania, New York, Wisconsin, and Virginia. He was previously employed at International Paper Company (Allegheny Timberlands) in Coudersport, Pennsylvania, where he was responsible for timberland management operations in north-central Pennsylvania.

**Lora A. Lucero** is a land-use attorney and city planner with a consulting practice in Albuquerque, New Mexico. She is editor of *Planning & Environmental Law*, a monthly publication of the American Planning Association, and is staff liaison to the American Planning Association's Amicus Curiae Committee.

**Curt Meine** serves as senior fellow with the Aldo Leopold Foundation, research associate with the International Crane Foundation (both located in Baraboo, Wisconsin), and director for conservation biology and history with the Chicago-based Center for Humans and Nature. He has authored and edited several books, including *Aldo Leopold: His Life and Work* (1988) and *Correction Lines: Essays on Land, Leopold, and Conservation* (2004). He is a past member of the board of governors of the Society for Conservation Biology and sits on the editorial boards of the journals *Conservation Biology* and *Environmental Ethics*. He is also active in local conservation as a founder and member of the Sauk Prairie Conservation Alliance in Sauk County, Wisconsin.

**Luther Propst** cofounded and directs the Sonoran Institute, with offices in Arizona, Montana, Colorado, Wyoming, and Baja California. The institute's mission is to inspire and enable community decisions and public policies that respect the land and people of the West. The institute works throughout the West on policies to improve the management of state trust lands, to better integrate conservation into land development, and to assist cities and counties to better manage growth.

**J. B. Ruhl** is the Matthews & Hawkins Professor of Property at Florida State University College of Law, where he teaches courses on environmental law, land use, and property. He is coauthor of *The Law of Biodiversity and Ecosystem Management*, the first casebook to organize environmental law under those emerging themes, and of *The Law and Policy of Ecosystem Services*, the first book-length treatment of the integration of ecosystem services into law and policy. Prior to entering full-time law teaching, he was a partner in the law firm of Fulbright & Jaworski, LLP, practicing environmental and natural resources law in the firm's Austin, Texas, office.

**John Shepard** is deputy director of strategic and program advancement at the Sonoran Institute. He launched the institute's training programs on community land-use planning for rural western county commissioners and partnership building for public land managers and gateway communities. He also established Building from the Best of Tucson, a Sonoran Institute project that promotes development that is consistent with Tucson's building traditions and appropriate for a desert community.

**Lynne Sherrod** is currently the western policy manager of the Land Trust Alliance, the national umbrella organization for land trusts across the country. She works on legislative issues related to land conservation in the West. Previous to that she was the first executive director for the Colorado Cattlemen's Agricultural Land Trust for nine years. She is a fifth-generation Colorado rancher and with her husband, Delbert, ranches near Collbran, Colorado.

**Diane Daggett Snyder** is vice president of community development for the U.S. Endowment for Forestry and Communities, a nonprofit corporation endowed to support educational, charitable, public interest, and forest sustainability projects in forest-reliant communities. Prior to that she served for over a decade as the founding executive director of Wallowa Resources, a local nonprofit organization that is promoting community development and sustainable management in the natural resources sector. She grew up in ranching as a fourth-generation resident of Wallowa County and managed her own herd on the family ranch for nearly twenty years. In 2007, *Oregon Business* magazine highlighted her work as one of fifty Great Leaders in Oregon, and she was the recipient of the Donald M. Kerr Award from the High Desert Museum.

**Courtney White** is a former archaeologist and Sierra Club activist who dropped out of the "conflict industry" in 1997 to cofound the Quivira Coalition, a nonprofit organization dedicated to building bridges between ranchers, conservationists, public land managers, scientists, and others. His book, *Revolution on the Range: The Rise of a New Ranch in the American West,* was published by Island Press in 2008. He lives in Santa Fe, New Mexico, with his family and a backyard full of chickens.

# INDEX